Liars, cheats and copycats

JAMES O'HANLON has travelled around Australia and the globe uncovering the secret lives of insects and spiders. He has published more than 30 academic papers, and his popular science writing has appeared in *ABC News*, *Australian Geographic*, *The Guardian* and the *Sydney Morning Herald*. He is the author of *Silk & Venom: The incredible lives of spiders*.

Liars, cheats and copycats

Trickery and deception in nature

JAMES O'HANLON

NEWSOUTH

UNSW Press acknowledges the Bedegal people, the Traditional Owners of the unceded territory on which the Randwick and Kensington campuses of UNSW are situated, and recognises the continuing connection to Country and culture. We pay our respects to Bedegal Elders past and present.

A NewSouth book

Published by
NewSouth Publishing
University of New South Wales Press Ltd
University of New South Wales
Sydney NSW 2052
AUSTRALIA
https://unsw.press/

Our authorised representative in the EU for product safety is
Mare Nostrum Group B.V., Mauritskade 21D, 1091 GC
Amsterdam, The Netherlands (gpsr@mare-nostrum.co.uk).

A catalogue record for this
book is available from the
National Library of Australia

ISBN   9781761170171 (paperback)
          9781761179280 (ebook)
          9781761178467 (ePDF)

Cover design Madeleine Kane
Cover images bird: rawpixel; fish: rawpixel; leaf insect: istock/The
    Nature Notes; rafflesia flower: istock /Elena Malgina; moth: istock/
    THEPALMER; Bengal tiger: alamy/The Natural History Museum;
    lizard: shutterstock/Ekaterina Gerasimchuk; moth: rawpixel; zebra:
    istock/The Nature Notes; octopus: rawpixel; lizard: rawpixel
Internal design Josephine Pajor-Markus

Contents

Introduction

As an enthusiastic young student of natural history, I had what I thought at the time was a fantastic idea: I was going to pack my bags and head into the jungles of Malaysia in search of one of the world's most beautiful insects, the orchid mantis (*Hymenopus coronatus*). To be fair, a few people had warned me beforehand that this was a silly idea. But having just spent a few years cutting my teeth in the science world as a student studying Australian praying mantises I thought, *how hard could it be?* It turned out to be very hard.

Orchid mantises are unique among all other praying mantises. They don't hide their charms away from the world, like the stoic brown or green praying mantises you might have seen before. Orchid mantises stand out like a shining beacon – they are bright white and pink, and their legs expand into large petal-shaped lobes. It's no surprise where they get the name orchid mantis; at a glance, they are an astounding simile of a bright blossoming flower. Despite their stunning beauty, no one had ever studied orchid mantises in the wild, and I was determined to be the first.

So off I went to Malaysia, eagerly expecting to stroll casually through the forest happily spotting hordes of bright white and pink praying mantises shining brightly in the undergrowth. I soon learned why no one had ever studied orchid mantises in the wild before: they are frustratingly rare. Instead of staking my claim as a groundbreaking praying mantis expert, I spent an embarrassing amount of time walking through the rainforest spotting what looked like flowers and getting very excited, only to quickly realise that they were, in fact, flowers.

At one point, I was given the name of a local wildlife expert. He was Orang Asli, the Indigenous people of peninsular Malaysia, and had lived his entire life in the rainforest. He was an elder in a nearby kampung, or village, and if anyone knew how to find orchid mantises in the wild, it was him.

I visited him a few times in his village, which was a small strip of wooden houses built into a small valley in an undulating mountain range. Each house in the village was a small square hut raised on precariously thin wooden stilts. The roofs were either made of corrugated metal sheets or woven from palm tree leaves. A collection of motorcycles lined up at the entrance of the village and a few scattered satellite dishes sticking from the roofs were reminders that I was still in a modern society, just one that lived a traditional lifestyle.

Each time I visited, we chatted in broken English. The few Malay phrases I had picked up along the way didn't help much and for some reason my pocket-phrasebook didn't include essential conversation starters like *what time of year are praying mantises most abundant?* He knew of orchid mantises and where to find them in the forest, but shared the same doubts as others who had warned me about going on this journey. I was probably not going to find them in the kind of numbers that would let me do field experiments. The overall conclusion was that finding wild orchid mantises was simply going to be a matter of persistence and luck.

Despite not learning any hidden secrets about finding orchid mantises in the wild, I learned many other things about life in the rainforest; like the fact that the local delicacy durian was considered a 'heaty' fruit and was so rich and flavoursome that it made you feel like your skin was warming as you ate it. I learned that some Orang Asli still hunt with long bamboo blow darts, and that wild tigers could still be found roaming in these forests. Intrigued, I asked the village elder if he had ever seen tigers in the wild. He said that he had, and held up three fingers. In an entire lifetime spent in the rainforest, he had seen tigers three times. This I found terribly exciting. I wanted to know more about finding tigers in the wild and kept asking questions. But my excitement was met with a stern glare. He quickly dispelled my enthusiasm and reminded me of the reality of stumbling across a tiger in the wild. It was not the exciting nature-watching experience that I, as a casual rainforest visitor, imagined it to be. Each encounter, for him, was a near-death experience.

The harsh reality of life in the wilderness hadn't fully occurred to me. The village was only a few hours' drive from a major city, and along a main road to tourist hotspots. Since Malayan tigers are incredibly rare and endangered, I had assumed they must all be hidden away somewhere far from civilisation. However, I learned that in this area, locals would often find tiger tracks in wet soil, scats on roadsides, and hear faint roars echoing through the mountains at night, but actual tiger sightings were incredibly rare. What was especially haunting about this man's stories was his awareness that, although he hardly ever saw tigers, the tigers *always* saw him.

It made me wonder, how is this possible? Tigers are bright orange and white with black stripes. Fully grown males are over 2 metres long and weigh over 130 kilograms. Even if they move stealthily, how on earth can these bright orange stripy beasts go unseen in the bright green settings of the Malaysian rainforest? It seems counterintuitive, but tigers are wonderfully camouflaged. In the rainforest, these enormous striped predators simply disappear among the complex of tree trunks, shrubs and shadows that make up the dense understorey.

For people like me, who have only ever seen tigers in the zoo, it's hard to imagine how this works. Animal camouflage in action is a type of deception – an optical illusion that overrides other animal senses and tricks them into thinking that there is nothing of interest in their field of view. The orchid mantises that I was looking for are another example – with a few simple modifications to their body shape and colour, these animals cease to be insects and are disguised as something completely different.

There are seemingly endless ways that animals and plants can deceive, cheat and swindle their way to survival. There are orchids that trick male wasps into collecting their pollen with the false promise of romance, octopuses that can change colour in an instant to match their background and fireflies that can flash to either lure a mate or an easy meal.

I've been fascinated by how living things have evolved to survive using trickery and deception. Diving into the science of deception can sometimes feel like shaking a tether that holds you down to reality. It reminds you that you can never quite trust your senses, or your instincts. Our eyes, ears and noses, as wonderful as they are, only give us a narrow sliver of information about the world around us. Right within our reach are unseeable worlds of light, sounds and smell that we are simply unable to perceive. Beyond those are worlds of sensory information that humans don't even have the organs to give us a small window of insight. Even where our senses can detect the world around us, examining deception in nature reminds us how vulnerable those senses are to misinformation and misdirection. When something as simple as a subtle colour gradient can decide whether our eyes and minds perceive a live animal right in front of us, it makes me wonder what other forms of deception are happening that we aren't even aware of. And if this deception occurs in sensory worlds beyond our own, is it possible for us to ever be aware of it?

In this book, we'll explore the many dastardly ways that plants and animals deceive and how research into deception is changing our understanding of how animal eyes, ears, noses and brains work. To begin, let's explore the world of animal camouflage. How do creatures disappear right before our eyes, and how can a giant orange cat hide away in a dark green jungle? I hope that reading this book pulls back the curtains and gives you a glimpse into worlds beyond your eyes and minds. I'll ask you to imagine yourself experiencing the world through the perspective of other living things. Then, once you have glimpsed their world, remind yourselves that nothing is ever as it seems.

1

For my first trick, watch this animal ... disappear!

As a science writer, I am supposed to start these chapters bragging about how many hours I have spent trudging through lush jungles, finding swarms of exotic-sounding insects. But if I am going to be completely honest, I've spent an embarrassing amount of time walking through forests not finding anything at all. Despite what some might claim, it's not because I'm a lousy field biologist, it's because most animals simply don't want to be found. In the battle for survival, being inconspicuous is a very good idea. It means avoiding unwanted attention from both predators *and* overenthusiastic field biologists.

The simplest way for animals to avoid being seen is simply to hide. Turn over any rock in your garden, and you're likely to disturb an animal that was, until you came along, doing a very good job of being inconspicuous. While many animals do very well at hiding in obvious places, like under rocks, in crevices and inside burrows, others find much stranger places to hide. Some small frogs, for example, will take refuge inside burrows that don't belong to the frogs, but to tarantulas. These large spiders dig holes into soil that become their permanent homes and, in some cases, they seem happy to share these homes with small frogs. The benefits for the frogs are obvious – they get a safe burrow with a live-in bodyguard. Whether tarantulas also benefit from this arrangement is unclear, especially since small frogs make a tasty meal for large tarantulas. North American narrow-mouthed toads (*Gastrophryne* spp.) secrete toxic chemicals from their skin that seem to deter tarantula attacks, so perhaps the spiders are begrudgingly tolerant of their froggy flatmates.

The award for most creative hiding place in the animal kingdom should probably go to the small crabs who take refuge inside the anuses of sea cucumbers. Sea cucumbers, if you haven't come across one before, are the ugly cousins of the starfish family. They look a bit like leathery sausages that inch slowly along the seabed. At the rear end of a sea cucumber is a small hole, which opens into a multi-functional space that also houses their gills. By opening and closing their anus, they pump water in and out, which expels wastes but also flushes oxygenated water past their gills. The branching cavities inside this strange anus/lung structure apparently make an ideal refuge for small crabs that live symbiotically inside sea cucumbers. As the sea cucumber pumps water in and out, it brings fresh water and an ongoing supply of plankton and organic matter that the crabs can feed on.

While hiding is a simple and elegant solution to avoid predation, for many creatures, spending their lives under a rock or inside something else's butthole isn't ideal. This is where camouflage comes in handy; it allows animals to live out in the open, free of the constraints of a burrow or crevice. At first, animal camouflage seems like it would be a very easy concept to understand – if an animal matches its background, it's hard to see. Simple, right? But when you think about it a little bit more, animal camouflage seems a bit absurd. Take an animal like the European nightjar, for example. These birds are often pointed to as classic examples of camouflage. They nest on the ground, making them and their eggs incredibly vulnerable to predators. Their main defence is their mottled plumage which makes them effectively disappear among leaves, twigs and soil. If you get the chance to look at them up close, however, they look like, well, birds. They have all the features you would expect a bird to have; like beaks, eyes, legs, wings and feathers. Yet somehow all these very bird-ish features combine in a way to make them *not* look like a bird.

What sort of optical illusions come into play when even animals, like humans, with very acute vision can be tricked so easily? It turns out that effective animal camouflage requires some sophisticated adaptations that hack the circuits of other animals' eyes and minds. And scientists are only just beginning to grapple seriously with how these complex strategies work to make things disappear right in front of our eyes.

Camouflage and colour change

Let's face it, when we say the word 'camouflage' we all know what we're talking about, right? You might picture a brown frog sitting on a brown rock or perhaps army personnel hiding in the bushes, wearing patchy green clothes and a few jaunty twigs poking out from their helmets. Perhaps the simplest way to describe camouflage would be to say that something blends in with its background. But say the word camouflage to a sensory ecologist – a scientist who studies animal senses and how they are used in the wild – and they will ask, *what type of animal camouflage? Is it crypsis or masquerade? Does it deter detection or recognition? Is it object imitation or element imitation?* And so on and so forth …

Of all the fields of natural science that I have come across, sensory ecology is overflowing with pedantry about different terminologies, categories and subcategories for different biological phenomena. For every paper published on animal camouflage, it feels like there are another two papers published to discuss the semantic implications of the terms the first paper used. While it's very important for scientists to have these discussions to make sure that they are being accurate in what they are describing, we won't be diving into the philosophical intricacies of the different terminologies here. If you are keen on diving into the semantic implications of *object imitation* versus *element imitation*, then by all means head to the reference list at the end of this book and find the research papers that tickle your fancy. But for now, let's get back to exploring how camouflage works in the wild.

The simplest way for a creature to camouflage is to match closely the colour of its background. This easily explains many of the patterns we see in nature, such as why we tend to find bright white Arctic foxes and hares scurrying around the snowfields of the Arctic Circle, and not in verdant green rainforests. Of course, such straightforward background-matching strategies quickly become hampered when the background changes; like, for example, when the seasons change and all the snow melts. Many alpine and polar animals, like snowshoe hares (*Lepus americanus*), alpine foxes (*Vulpes lagopus*) and rock ptarmigans (*Lagopus muta*), have seasonal moults so that during the winter their coats are bright white, but gradually turn greyish brown for the warmer months.

Many animals have ways of changing colour to assist with camouflage in different environments. Small crabs, such as the horned ghost crab (*Ocypode ceratophthalmus*), that run around the beaches of South-East Asia, change their colour from dark to light depending on the brightness of the sand underneath them. They can go from being milky white to mottled black within the space of a few hours, through the movement of pigment granules inside their exoskeletons. Not only can they change to better match the sand beneath them, they also change colour based on circadian rhythms, turning lighter during the day, and darker at night.

Chameleons are some of the most famous colour-changing animals. Despite what cartoons might have taught us, they can't turn invisible by changing into an infinite repertoire of technicolour patterns. In reality, their colour change is slightly more subtle. South African dwarf chameleons (*Bradypodion* spp.), for example, can change between green and brown to match the plant parts they are resting against. They are able to change colour within a matter of seconds by moving around different layers of pigment cells and reflective nanocrystals imbedded in their skin.

Where animals can't change colour to match their background, just picking the right place to hide can go a long way to maximising their camouflage. There are many examples of animals known to pick their resting places to better match their own colour. Pacific tree frogs (*Pseudacris regilla*) come in either brown or green, and actively choose to sit on the brown or green backgrounds that best match their own colour. There are many moth species with mottled and patterned wings that align their bodies so that the patterns on tree trunk bark matches closely their own colour and patterning. It's likely that the moths can feel the texture of the bark and use this information to select the best hiding spot. There are even examples of birds that pick nest sites that will match the mottled colouration of their eggs. The mottled eggs of the Japanese quail (*Coturnix japonica*) vary from mostly white with a few dark specks, to almost entirely black. They lay their eggs in nests dug into soil and, before laying, females somehow select nesting sites with soil that matches the colour and mottling of their eggs. The eggs are then harder for predators to find, should the mother need to leave the nest.

These stories raise befuddling questions about animal psychology. It seems to suggest that animals comprehend their own colouration or have some understanding about what camouflage is and how it works. It's easy to imagine something like a snake turning its head, observing its own body with its large acute eyes, and assessing how closely it matches its background. This is harder to imagine in the case of a moth whose small compound eyes can't see its own wings at rest, or a bird that picks an appropriately camouflaged nest site before its eggs have even been laid.

What biological hardwiring is going on inside these animals that lets them pick the appropriate hiding place without the ability to actively observe how closely they match the background? Scientists are still only scratching the surface of how animals make decisions that will maximise their camouflage.

When it comes to understanding how animal behaviour and mindset play a role in camouflage, there is one group above all that challenges our preconceptions about animal capabilities. And as impressive as the aforementioned feats of colour change are, this group of animals outpaces all others for their colour-changing wizardry. They are the incredible cephalopods.

Aquatic quick-draw colour change

You may be more familiar with them being called octopuses, squid or cuttlefish. Cephalopods don't need to wait for seasons or even minutes to pass for their colour to change; they can, quite literally, change colour at the speed of thought.

If you were to look up close at the skin of an octopus, you would see that it is covered with small colourful dots. They can be yellow, orange, red, brown or black. From a distance, the individual dots are imperceptible and all blend together. So, you perceive the colour of the octopus just like the colours you are seeing now on the screen or printed page you are reading – tiny individual pixels combining to create an illusion of shape and pattern. Each dot on the octopus's skin is a small organ called a chromatophore that is made up of a tiny elastic sac filled with pigment. Small muscles around each sac pull it open, exposing the dark pigments within. Unlike some other colour-changing animals who must wait for hormones to kick into action for their colours to change, a cephalopod's chromatophore muscles are switched on and off instantaneously by nerve impulses. Individual nerve fibres run directly from the brain to small groups of chromatophores and can be controlled independently.

If you were to once again look closely at an octopus's skin while it is changing colour, you would see these colourful dots growing and shrinking as the chromatophore muscles flex and relax. With the power of this complex network, the colour of a cephalopod needn't be uniform – they can generate complex patterns with contrasting dark and light patches. Colour and pattern changes can happen in a matter of milliseconds. To control and coordinate these colour-changing organs, cephalopods need big brains. In the common octopus (*Octopus vulgaris*), the part of the brain responsible for controlling the chromatophores contains over half a million brain cells. It's been suggested that cephalopod intelligence is a by-product of needing large brains to coordinate their network of chromatophores.

What this all means is that cephalopods have a system whereby they can produce continuously variable camouflage. Scientists from the University of Chicago have been working for decades to understand camouflage in the common cuttlefish *Sepia officinalis*. When these cuttlefish are viewed against a uniform sandy background, they take on a uniform sandy colouration. When the sandy background is mottled with larger particles, like scattered rocks and seashells, the cuttlefish display a mottled colour pattern with a smattering of dark spots. When the background is even more complex and comprised of large pebbles, the cuttlefish changes its patterning to large patches of contrasting colours. Even when the cuttlefish are tested in the laboratory and placed against a completely artificial background (e.g. a black and white checkerboard pattern), they adjust their camouflage to fit the size of the checkerboard squares on the ground below them. When the checkerboard pattern is very fine scale, the cuttlefish show fine-scale mottling, and when the checkerboard pattern is large, the cuttlefish display large contrasting colour patches. They are somehow able to assess the complexity of the environment around them and adjust their camouflage accordingly.

In a sense, cephalopod skin is like a computer-controlled screen, where each chromatophore is a pixel hardwired directly into a processor, the cephalopod brain. This gives them an incredible amount of direct control when changing colours, allowing them to customise their camouflage on the spot depending on the background. And if this wasn't hard enough to comprehend already, as far as we can tell, cephalopods are colourblind, and manage to do this without being able to see their own colour match to the environment the way we or other animals can. So how on earth do they manage such masterful camouflage?

First, they don't need to see their own colour to adjust their camouflage. In an ingenious (and adorable) experiment, scientists put little collars on cuttlefish so that they couldn't see their own bodies. Think of the collar that vets put on dogs after an operation so they can't lick their wounds, then imagine a miniature one of those on a cute little cuttlefish. Even when these cuttlefish couldn't see their own bodies, they adjusted their camouflage to match their background just as well as they could without collars on.

Secondly, colour matching might not be as important as it first seems, and cephalopods can get most of the way towards camouflage simply by matching the overall brightness of their backgrounds. Scientists have examined this using the checkerboard method mentioned above. Instead of using a pattern with black and white squares, the scientists created a checkerboard with squares differing in colour, but not overall brightness (e.g. bright blue and yellow squares). When the cuttlefish were put into these enclosures, they were completely unable to detect the colour complexity of their background and adopted a uniform colouration that matched the brightness of the background, rather than a mottled pattern as they did against black and white squares.

Finally, chromatophores aren't the only trick that cephalopods have up their sleeves. Embedded in their skin are other microscopic structures, like leucophores that lie underneath the chromatophores, which, when exposed, scatter diffuse light reflecting the ambient light in the environment. So, a cuttlefish hidden among some green algae can acquire a greenish tinge by reflecting some of the ambient green light from the adjacent algae. Some species have complex reflective structures in their skin (iridophores and reflector cells) that scatter additional blue-green light from the surrounding environment. By using all of these in combination, cephalopods can put together a variable suite of camouflage strategies even without having their own colour vision.

Of these three groups of cephalopods, squid are perhaps the humblest when it comes to colour change. They have comparatively fewer chromatophores on their skin and, as squid tend to swim about in open waters, they are less famous for their flashy camouflage compared to octopus and cuttlefish that usually hang around rocks and algae on the seafloor. Octopus and cuttlefish can be absolutely covered in small, densely packed chromatophores.

They don't just stop at changing colour either. They take camouflage one step further, by changing the texture of their skin. Along with chromatophores and the like, octopus and cuttlefish have groups of muscles in their skin that, when activated, project outwards to form dull spikes and flattened lobes. These undulating structures, called papillae, can be adjusted to match the roughness or surface complexity of the environment. So, when a cuttlefish hides within a clump of rough seaweed, it will extend its papillae outwards, giving its skin a rough leafy texture. When the same cuttlefish hides against flat featureless sand, it can retract its papillae, leaving its skin silky smooth. The tentacles can come into play here as well. In a laboratory study, cuttlefish were placed in enclosures with walls covered in either horizontal, vertical or diagonal stripes. At rest, the cuttlefish would hold their tentacles out in alignment with the stripes on the walls around them. Similar behaviour has been observed in the wild, where cuttlefish will orient their short tentacles to align with nearby sea grasses and coral branches.

But let's face it, apart from some showy octopus antics, animals camouflaging by blending in with their background is a straightforward concept, right? Animals that live among green grass are a grassy green, those that live among brown sand are a sandy brown, and those that live in the dark blue ocean are, you guessed it ... bright orange. Makes perfect sense, no? Well, it turns out that background matching isn't as straightforward as we might think. The examples of background matching we are most familiar with are ones that we can appreciate with our human eyes, and in environments that we are comfortable living in. But there are many other strange animal eyes out there, and unfamiliar environments that make background colourmatching a little less straightforward than meets our eyes.

Let's dive deeper into how this works by diving back into the ocean.

Colours are an illusion

I learned to scuba dive when I was a teenager. My uncle, Ed, was a dive instructor, and as soon as I was old enough, he took me out into the cold waters of Sydney Harbour and the south coast of New South Wales, and introduced me to the beauty and wonder of the southern seas. These aren't the familiar coral reefs that you see in dive brochures. These are underwater forests thick with sea grasses. The water is often bitterly cold, but worth adventuring in for the chance to spot enormous blue gropers, majestic weedy sea dragons and beautifully frilled wobbegong sharks.

On one trip, we went down to Kiama and dived the sunken wreck of the SS *Bombo*. To get there, we took a boat out to sea and jumped in the water immediately above the site of the wreck. As I bobbed about on the surface of the water, there was nothing to see but chopping waves and grey clouds. I find that boat dives like this bring about strange feelings; parts calm serenity of the featureless open ocean, parts fear of the unknown darkness below. Once the whole dive team was in the water and ready, we began our descent. The second my eyes dipped below the surface, I entered another world. There was nothing around but cool blue on all sides. And below, nothing but blackness. As we descended further, the blue water became darker and darker, until we eventually came to rest on the grey sand below. The hull of the *Bombo* arched above us like a dark monolith, with jagged outlines warped by decades of rust and encrusting barnacles.

Once the team settled on the ocean floor, Ed signalled to us all to pay attention to what he was doing. He reached into a pocket on his dive vest and pulled out a small piece of dark blue plastic. Like a magician introducing a magic trick, he held it out to each of us in turn, giving us a good look at this innocuous piece of blue plastic. Then, with his other hand, he pulled out a small dive torch. He pointed it at the piece of blue plastic and, when he flicked the switch, a vivid red circle appeared in the centre of the piece of blue plastic. With only natural light at this depth, you would have sworn that the piece of plastic was blue. Only by using torchlight could we make out its bright red colour.

Ed then took his torch and pointed it at the sand below us – it wasn't a murky grey like I thought, it was bright sandy orange with flecks of colourful shells and rocks. The hull of the sunken ship, when illuminated by the torch beam, was encrusted with a rainbow of pink, yellow and purple sponges. As we descended into the water's depths, the world around us wasn't just getting darker, it was becoming bluer or, more precisely, less red.

Water absorbs light, and the deeper you go, the more light it absorbs and it doesn't absorb all wavelengths of light equally. Let's take a quick detour and talk about the physics of light and this strange concept we call 'colour'. Light is electromagnetic energy that vibrates at different frequencies. Colour is how our eyes and brains decipher different wavelengths of light. Our eyes can detect light at wavelengths between around 380 to 700 nanometres (nm). At the high wavelength end of the visual spectrum, we have red light, and as we move down in wavelengths, we pass through all the colours of the rainbow; oranges, yellows and greens, until finally we get to the low wavelength colours of blues and violets. Beyond these wavelengths are other forms of electromagnetic radiation that our eyes can't see. Above 700nm are infrared radiation, microwaves and radio waves. Below 380nm are ultra-violet (UV) light, X-rays and Gamma rays. Animal eyes differ wildly in the wavelengths that they perceive – a topic we'll dive into a bit later. Many animals can see UV light and others can see into the low infrared end of the spectrum, but beyond these wavelengths, electromagnetic energy is considered non-visible.

Luckily for us earthlings, we have a readily available source of visible light energy: the sun. The light from the sun shines approximately evenly across wavelengths of visible light. So when you walk out into your garden and look at some nice green leaves, what is happening is that light energy particles (photons) shine down from the sun, they bounce off the surface of the leaves and are intercepted by our eyes. Most objects don't reflect light back evenly. Those green leaves are reflecting green light at somewhere around 530nm, while all the other wavelengths of light are absorbed at the leaves' surface. Similarly, yellow flowers will reflect yellow light at around 580nm, and red tomatoes will reflect red light somewhere towards 700nm. Purple objects might be reflecting a combination of both blue and red light. White is made from all colours of the rainbow, so things that appear white to us do so because they *reflect* light from across the entire visible spectrum. On the other hand, things that appear black *absorb* light from across the entire spectrum. Through light reflecting in different combinations across the visible spectrum, we get all the myriad colours of nature.

As you descend into the ocean, the water molecules absorb light, starting with low energy reds and oranges first.* You only need to descend about 10 metres until all traces of red and orange light have been absorbed. As you go further down, yellows and warm greens disappear, and by the time you get to 50 metres, the only colour left is a cool blue at a wavelength of around 480nm.

---

*   It can get confusing because low-energy frequencies have high-wavelength values. The wavelength value is the distance in nanometres between peaks of the photon wave or the speed at which light particles wobble about. Red light photons vibrate slower, so have less energy than blue light. This means red light has a greater distance between vibrations, thus a higher wavelength value (~700nm) than blue light (~500nm).

This means that at different depths in the ocean, animals live in vastly different lighting environments and need to use different camouflage strategies depending on their habitat. In surface waters and within shallow reefs, the lighting conditions are like those on land, so background matching can look like what we would expect it to. Away from the coast, things start to get a bit stranger. In the open ocean, there isn't much to hide against and the background that these animals must camouflage against is the murky blue of the sea. It's not always murky blue of course; the colour of the background can change entirely depending on things like the time of day, weather conditions and water turbidity.

Open-water animals have evolved clever ways of hiding, even when floating around in completely transparent water. One solution is simply to be transparent as well. Many small jellyfish, comb jellies and salps are see-through. Light passes through their tissues instead of bouncing off their surfaces. They are far from being perfectly invisible, and under close inspection you can still make out their different forms and shapes. But this level of transparency may just be enough to escape the attention of animals at a distance or ones that have poor sight.

Of course, perfect invisibility would be the ideal camouflage solution for all animals, but it turns out to be very difficult. It doesn't just require that the animal's skin be transparent, but all the animal's insides must be transparent, too. Animals like jellyfish have relatively simple body plans, and most of their mass is made up of non-living jelly. Even their gelatinous insides refract light slightly differently to the water around them, making them transparent but not invisible. And while there are examples of transparent fish, squid and crustaceans, the general rule is that for larger and more complex animals, transparency simply isn't an option. Animal tissues and blood are packed with pigments that, as you no doubt know, are what give things their colour. Haemoglobin, for example, carries oxygen around in blood and just so happens to be a chemical pigment. If a large and complex animal needs a circulatory system to survive, then transparency is off the table as a camouflage strategy.

Some open-water fish have evolved the complete opposite strategy to transparency, which is to *reflect* as much light as possible. Many open-water fish have glistening silver scales that act as mirrors. The scales are aligned vertically and bounce back whatever light is in the ambient environment. In the open ocean, the murky blue on one side of the fish is roughly the same as the murky blue on the other side of the fish, so the mirror-like scales reflect a colour that matches whatever the fish's background is.

You can see this in action yourself if you ever come across these types of fish when diving or snorkelling. You might be able to see from below the shadowy form of a fish swimming up near the surface. As you come up to eye level, certain scales start to reflect light from the right angles. It almost looks like holes appear in the side of the fish and you can see right through them into the blue sea beyond. Then, when you see them from side on, at just the right angle, well ... you don't see them at all! They disappear entirely but for a vague outline and maybe a big eye looking back at you. It's only when they dart away that the mirrors flicker and the fish is revealed.

This is a literal magic trick found in nature. If you have ever watched a magician make their assistant disappear from on top of a table or perhaps saw them in half, chances are you have seen this mirror illusion in action. From a distance, it looks like there is empty space underneath the table where you can see the floor and some red velvet curtains. In fact, this empty space is one or more mirrors that reflect part of the stage off to the side, which also happens to include a view of some very similar-looking floor panels and velvet curtains. The disappearing magician's assistant, or parts thereof, are simply hidden behind the mirrors underneath the table. If a simple parlour trick like this can hide an entire human on stage, then it's no stretch of the imagination to see how a covering of mirror-like scales can make a fish disappear in the open ocean.

Many open-ocean fish do simply have a murky blue colour to blend in with their murky blue backgrounds. However, as you travel deeper into the ocean, the light environment begins to change, and this is where background matching starts to become a little counterintuitive. Once you get below 10 metres in depth, all of the long wavelength red and orange sunlight has been absorbed by the water above, and this is where you start to see deep-sea creatures with reddish pigmentation.

The deeper you go, the more red creatures you will find; from schools of orange roughy, to crimson-red vampire squid. The reason for this is simple – in an environment where no red light exists, the colour red doesn't exist. The same goes for other long wavelength colours, like oranges and yellows. On our dive trip to the wreck of the *Bombo*, when my uncle pulled out the small piece of plastic, it was blue because only blue light was present in the environment. It was only when Ed pulled out the torch and introduced artificial light to the environment, a light that covered the whole visible spectrum, that other wavelengths were present to bounce off the underwater objects. The reason we describe those deep-sea creatures as red is because we go underwater with torches, or in deep-sea submersibles with great big spotlights on the front of them. Or we haul up creatures from the ocean depths and look at them on the surface in full-spectrum lighting. But this doesn't reflect what they look like in nature because where they live, red light, and thus red colours, simply don't exist.

As you travel deeper into the ocean, you eventually reach a point where all downwelling light has been absorbed. At about 1000 metres in depth, you reach the mysterious-sounding 'midnight zone' where there is no trace of sunlight at all, and creatures live in pure darkness. This environment is so different to our own that life tends to take on bizarre forms unlike anything we are used to seeing in the shallower ocean. This is the realm of the fabled blob fish, sharp-toothed anglerfish and colossal squid. You might assume that, since there is no sunlight, camouflage wouldn't matter, and animals could be free to evolve all the colours of the rainbow. But, while there is no sunlight here, there can still be light from other sources.

Not content with just looking strange, some deep-sea creatures evolve bizarre adaptations to survive in the dark ocean depths, including having built-in bioluminescent searchlights. Perhaps the most epic example of this is the colossal squid (*Mesonychoteuthis hamiltoni*). This animal lives up to its name; it's the largest invertebrate on the planet and has the largest eyes in the animal kingdom. Fully grown, they can weigh close to a half a tonne and, with their tentacles fully stretched, reach up to 6 metres in length. They are top predators and it's possible that they manage to catch prey using sight, even in the blackness of the deep. Around the edges of their eyes are large oval-shaped organs that house bioluminescent bacteria.

Keep in mind that no one has ever seen colossal squid hunting in the wild. As is true with many deep-sea creatures that we only know of from caught specimens, no one has ever actually observed how they catch prey. But looking at this combination of organs, it's possible that they are illuminating the depths with bioluminescent light, and hunting for prey using their super-sensitive soccer-ball-sized eyes. Many deep-sea animals use bioluminescence to create their own light in the ocean depths. Just like the colossal squid, they have specialised organs that house special bacteria that produce light as a by-product of their metabolism. Often, the light produced by these bacteria glows at around 480nm, which we have already learned is the colour that travels furthest through water. By using this coloured light, in combination with eyes that are very sensitive to this particular wavelength, predators may still hunt using vision in a place where sunlight never reaches.

How, then, do animals camouflage from predators with built-in search lights? Scientists have discovered a range of fish that have ways of blending into the pure blackness of the midnight zone. They aren't just black; they are *ultra black*. They are blacker than black. Blacker than a Spinal Tap album cover. Look around you and find something nearby that's black. Have a good look at it and, adopting your inner metal-head, think: *is it really black? Could it perhaps be blacker?* Right now, I'm looking at my laptop, which is a matte black colour. But if I'm honest, it's more of a very dark grey. There are parts that shimmer white where the light hits it at the right angle, and the more I look at it, I wonder if I can see a bit of a reddish tinge to it.

For something to be pure black, it must absorb as much light as physically possible. Animals like these deep-sea fishes are described as being ultra-black when they absorb over 99.5 per cent of all light. This is incredibly rare in nature and, even in synthetic objects, achieving ultra black poses some serious chemical engineering challenges. The deep-sea anglerfish (*Oneirodes* sp.) holds the record for having the blackest surface in nature, absorbing 99.956 per cent of all light. This and other ultra-black fish manage to do this by having dense packets of melanin throughout the surface of their skin, presumably to blend into the pure blackness of the deep ocean, even when glimpsed through the searchlights of a bioluminescent predator.

The ocean provides wonderful examples of how different lighting conditions can affect the way we think about colour, depending on the nature of light within the environment. The same can be said of different environments on land. Ambient lighting conditions change constantly with the time of day, weather conditions and aspects of the local environment. The bright directional light that shines down from the sun in open fields or desert plains creates a very different lighting environment to the diffuse or mottled light that makes its way through to the understorey of a dense canopied forest. The most obvious example of different light environments on land is the contrast between daytime and nighttime. Under the dim glow of moonlight and the even dimmer glow of starlight, camouflage is still necessary to avoid the attention of other nocturnal animals that have evolved super-sensitive night vision.

A team of scientists filmed the behaviour of giant Australian cuttlefish (*Sepia apama*) during mating season, using remotely operated vehicles. They found that during the day, cuttlefish were active, hunting and displaying mating behaviours. However, at night, the cuttlefish mostly ceased activity and settled on the seafloor, adopting camouflaged colour patterns. This strongly suggests that, even under the cloak of darkness, cuttlefish are still susceptible to visual predators and that they can modify their camouflage to suit nocturnal conditions. Since scientists are mostly humans, they tend to be most active during the day, and we know very little about how camouflage works at night, or what most nocturnal animals are up to in general. They are only just beginning to dive into how camouflage works in the darkness of night. This also raises a very interesting point about how camouflage could work when there is no light at all, or when it relies on senses other than vision, but that is a story for later.

When I was telling the story about how my uncle revealed to us the red colour of the seemingly blue piece of plastic, I was tempted to use phrases like the 'true colour of the plastic was revealed'. But is there even such a thing as a 'true colour'? You have no doubt heard the old philosophical thought experiment, 'If a tree falls in a forest and there is no one around to hear it, does it make a sound?' The answer is a little bit dull if we get pedantic about the definition of sounds. If 'sound' refers to the detection of vibrations in the air by an ear or another sensory apparatus, then, by definition, the answer is no. There can be no sound without a listener. The question of whether that same falling tree generates vibrations in the air molecules around it is another premise altogether. The same is true for colour. The word 'colour' refers to the way light is perceived by an eye and perhaps even the brain attached to that eye. So, the colour of something is entirely subjective, and depends on the light environment and what type of creature is looking at it.

To fully understand how camouflage works in nature, we must understand how other animals perceive the world. This adds even more complexity to the not-so-intuitive subject of background matching, and it starts to explain how a bright orange tiger can disappear within a forest of green leaves.

The eyes of the beholder

Some of you reading this would have spent a non-trivial amount of your childhood, like I did, with your face pressed up against the television, trying to make out the teeny-tiny red, green and blue rectangles that pixellated the screen. If this sounds familiar to you, it is a surefire indicator of our age, not only because it was at a time where people still had enormous cathode-ray tube televisions that took up half the lounge room, but also because it was a time before the internet and smartphones. Infinite forms of distraction weren't available to us twenty-four hours a day at the swipe of a finger, so kids were obliged to invent their own forms of entertainment, like going outside and seeing how long you could look at the sun, sticking your head in the sink and trying to keep your eyes open while you ran the tap over your face, or pressing your nose so close to the TV that you could see the individual RGB (red, green and blue) pixels that made up the image on screen. It's probably no surprise that we all need to wear glasses for reading now and definitely can't focus close enough to see the TV pixels anymore.

At the time, I didn't understand what I was seeing. From a distance, the image on the TV screen showed all the colours of the rainbow, but up close, with my nose pressed against the warm backlit glass, there were only three colours arranged in neat little rectangles – red, green and blue. I couldn't understand why there weren't yellows, pinks, whites and all the other tones I could make out from a distance. Like we covered above, colour is simply that which is perceived by the mind, and TV screens play a clever trick on us. They don't need to project the different colours of the rainbow; they just need to give us the right type of information for our brains to fill in the blanks. And it works the exact same way in modern projector screens as it did in your old cathode-ray tube televisions.

Our eyes have three different types of cells (photo-receptors) that are sensitive to three different wavelengths of light: red, green and blue. We experience the world in all colours of the rainbow because of the way our brains interpret the information we get from these three colour channels. Yellow light, for example, sits on the electro-magnetic spectrum between red and green light. So, when yellow light hits our eyes, it stimulates both our red and green photoreceptors roughly equally. Our brain puts together this partial information from red and green photoreceptors, and creates the colour yellow inside our minds, even though we don't have photoreceptors that are specifically tuned to detect yellow light. The same goes for all other colours of the rainbow; our brain has a clever system of interpreting partial information from our red, green and blue photoreceptors to let us experience many different colours. Televisions and other RGB-based projection systems hack into this system, making us think we are seeing colours that aren't really there. So the seemingly yellow object on the screen is actually a *greenish-red*, created by the combination of both red and green pixels.

Other animals can have completely different sets of photoreceptors, so they perceive the world in completely different ways. Cephalopods, as we mentioned before, are colourblind; they only have a single type of photoreceptor in their eyes sensitive to light somewhere around 490nm.* This means that they see the world in monochrome tones, based on variations in brightness of the bluish-green light that their eyes can detect. Other animals have photoreceptors that allow them to see colours that we can't. Ultraviolet light, for example, is easily perceived by many animals but completely invisible to us. Bee vision

---

* There is one known exception to this: no one knows why, but for some reason the firefly squid (*Watasenia scintillans*) has three types of photoreceptors.

has been studied extensively, and we know more about their visual systems than most other animals. Bees have three different types of photoreceptors: green, blue and ultraviolet. So, while they can perceive ultraviolet light, they can't detect red light. Some birds have four receptors: red, green, blue and ultraviolet. And, reaching up into the other end of the spectrum, some butterflies and leafhoppers have photoreceptors that can detect near-infrared light. There are myriad different photoreceptor types and combinations across the animal kingdom. Just as we put together an image of the world based on partial information from our three photoreceptors, other animals put together different images of the world depending on their different combinations of photoreceptors.

What all of this means is that if an animal's camouflage strategy has evolved to hide from something other than humans, then our eyes aren't the best judges of their camouflage. Which takes us, finally, to the case of the orange tiger in the green jungle. So far, we have been talking about animals using camouflage to hide from their predators, but it's just as useful for animals to hide from their own prey. Tigers are ambush predators that need to camouflage so that they can sneak up on unsuspecting prey, like deer and wild boar.

We see tigers as orange because the light that reflects from their coats spans the yellow and red parts of the visible spectrum. Their prey, however, have very different combinations of photoreceptors compared to humans. Most other mammals lack photoreceptors that can detect red light. Animals like deer and boar generally only have two types of photoreceptor in their eyes: ones sensitive to blue light, and ones sensitive to yellow-green light. From these animals' perspectives, the red light reflecting from a tiger's coat is imperceptible. This leaves only the yellowish-green hues that reflect from the tiger's coat, much like the yellowish-green hues of their jungle backgrounds. Scientists running computer simulations of animal vision have suggested that animals like deer and boar, which have only two photoreceptor types, are less likely to detect camouflaged animals like tigers based solely on colour. From a human perspective, we might imagine the orange coat of a tiger to be a strange adaptation for hiding in a green jungle. Once we adopt the perspective of another animal, the world starts to make a little more sense.

Our limited perceptions of animal colouration can mislead us in other directions as well. Crab spiders hide among flowers and ambush visiting insects as prey. They come in different floral colours, like white, yellow and pink, that help them blend in with the flower petals. Scientists have been studying this camouflage strategy in European crab spiders for decades, but once they turned their attention to Australian crab spiders, they had to start re-writing the crab spider rule book. While Australian crab spiders look bright white or yellow to us, when scientists measured their colour, they found that some species shine brightly in ultraviolet light. So, while it looks to us like Australian crab spiders blend in with their background, from the perspective of the bees that they are preying upon, the spiders stick out like a sore thumb. If you're wondering why ambushing spiders would evolve to look conspicuous to their prey, there is an answer. But I'll leave that juicy mystery for later in the book.

To study colour objectively, scientists need to find ways of taking their human biases out of the equation. We can use all manner of fancy lasers and cameras to measure the wavelengths of light that bounce off objects, including wavelengths we are unable to perceive. This way we can quantify something's colour, rather than simply lumping it into a descriptive colour category. Scientists can then consider the nature of the ambient light in the environment and build simulations of how other animal eyes might perceive these reflectance curves based on what we know about how different animal eyes work.

Now don't worry, I'm not going to insist we spend the rest of the book talking objectively about the 'reflectance spectra' of surfaces, or listing how reflective things are at particular wavelengths, under specific illumination, and as viewed by so-and-so's photoreceptors. For ease of reading, I will still just talk in general terms about the 'colour' of things, knowing that we are referring to our own biased perceptions. The take-home point is that animals sense the world very differently to how we do.

Throughout the book we will continue diving into strange sensory worlds that are invisible to us, but vivid in the senses of other animals. Understanding differences in colour is simply one example of this, and background colour matching is only one component of camouflage. Eyes can detect so much more than colour, and in the next chapter, we'll explore how animal camouflage plays around with shape and contrast to create some uncanny optical illusions.

2

The science and art
of camouflage

It's incredibly hard to imagine living in a world where scientific facts that we now take for granted were unknown. Camouflage and background matching seem intuitively obvious, but we should remember that we live in a post-Darwinian world where this sort of logical reasoning is already embedded in our collective psyche. If you haven't already, I would recommend going back and reading Charles Darwin's famous and world-changing book from 1859, *On the Origin of Species*. I'm not recommending it because you should expect a rollicking good read. Quite the opposite, actually. My immediate impression when first reading this classic book was how boring it all seemed. Take this quote, for example: 'When we see leaf-eating insects green, and bark-feeders mottled-grey; the alpine ptarmigan white in winter, the red-grouse the colour of heather, and the black-grouse that of peaty earth, we must believe that these tints are of service to these birds and insects in preserving them from danger.'*

---

* Charles Darwin (1859) *On the Origin of Species By Means of Natural Selection*, John Murray, p. 88.

With the curse of hindsight, Charles Darwin now seems passionately verbose in pointing out the bleeding obvious. This is how much of the book reads, but I still recommend that you read it because, after a while, you start to realise the importance of the fundamental statements that Darwin makes. It's a bit like listening to the Beatles as a teenager (or whatever band your parents tried to convince you was cool as a kid). You don't quite understand the hype because it's just old music that sounds like every other pop song on the planet. It's only with the passage of time and experience that you learn to appreciate the fundamental and world-changing nature of what you are hearing, and that the Beatles don't sound like everyone else, everyone *else* sounds like the Beatles. It took great minds like Darwin to articulate these fundamental principles so that they could later become common, assumed knowledge.

When it comes to understanding animal camouflage and colouration, there are other great minds who have influenced our thinking just as much as Charles Darwin. They are the Rolling Stones and Led Zeppelins of evolutionary biology. One such rockstar is Hugh Bamford Cott, who is often referred to as the father of animal camouflage research. In 1940, Cott published a book called *Adaptive Coloration in Animals*. To call it a book is a bit of an understatement – it's a 500-page, hardbound tome that you could use as a foundation for a small house. In this epic treatise, Cott lays out in laborious detail everything you could possibly want to know about why animals look the way that they do. The first half of the book covers animal camouflage, where he puts forward many ideas about camouflage that are still being studied by scientists today. He also outlines how an understanding of animal camouflage could be applied for military use, and takes the opportunity to criticise the British Army for failing to protect their troops by not adopting proper camouflage techniques. During World War II, Cott worked with British Army engineers to develop and deploy camouflaged ships, vehicles and weaponry.

Though many of Cott's ideas have stood the test of time, it's interesting to note that many of them didn't just stem from studies of the natural world, they were actually derived from an understanding of artistic principles. Cott, as well as being a seasoned naturalist, was also a skilled artist and photographer. Furthermore, his ideas about animal camouflage were heavily influenced by the work of another intellectual rockstar, American painter Abbott Handerson Thayer. While he was most famous for his incredibly realistic portraits, Thayer was also a passionate naturalist and created equally vivid paintings of wild animals and natural landscapes. In 1909, Thayer published his own immense hard-bound tome on animal camouflage called *Concealing-Coloration in the Animal Kingdom*, which formed the basis for much of Cott's later work.

It makes sense that Cott and Thayer, both trained in visual arts and natural sciences, would direct their attention towards how illusions work in nature. All visual arts are, in a sense, a form of optical illusion. Whenever you look at a famous painting, like *Girl with a Pearl Earring* by Johannes Vermeer, our brains instantly say, *oh look, a girl, and she has a pearl earring*. The reality is, though, that there is no girl, no pearl earring, no blue headdress, no golden shawl, just a bunch of paint smeared on a canvas. And yet at the hands of a skilled artist, this jumble of pigments is arranged in a way that convincingly dupes our brains into thinking that we are seeing a pale face turned in our direction and staring right into our eyes. An artist's skill lies in knowing how to use contrasts of light and dark to guide the human mind towards specific conclusions. They know how the placement of certain shapes, in a particular arrangement, is interpreted by our minds as a face. They know that stark contrasts of light and dark will draw our attention more than subtle tonal gradients, and that certain gradations of light and dark can give the illusion of three-dimensionality to a flat canvas surface. Understanding visual arts is, in very simple terms, knowing how to use colour, line, tone and texture to make viewers perceive objects that don't actually exist on the paper, canvas or screen in front of them.

What Cott and Thayer noticed when they were trying to paint and draw animals in their natural environments was how difficult it was to make the animals stand out as focal points in their paintings. They both lamented that drawing an animal accurately in its natural setting results in a work of art where the animal doesn't really stand out and look interesting like a 'proper' work of art should. Using their artistic sensibilities, they were able to articulate why this happened, and realised that wild animals seemed to be coloured and shaded in ways that counteracted the techniques an artist would use to make objects conspicuous in a work of art.

A simple example of how animals break the rules of classical art training is colour matching. To make an object stand out in a work of art, you create a contrast between the object's colour and the background. As we covered in the last chapter, camouflaged animals do the complete opposite. Important as colour matching may seem for animal camouflage, both Cott and Thayer deemed it to be of secondary importance to tonal differences. Contrasts between light and dark patches on animals seem to play a much more important role in camouflage than colour matching.

Visual artists know that contrasts and gradients between light and dark are key to realistically depicting three-dimensional objects, completely regardless of colour. If you decide to take up a drawing class at any point you should be prepared to spend weeks with a lead pencil shading spheres, cubes and cylinders in tones of light and dark before you ever think about picking up a coloured pencil. It is because our eyes and minds can make complete sense of the world using only tonal differences. Colour is often just a nice bonus. Think about how we can easily make sense of black and white films and photography despite the complete absence of colour. Think about how you can open a kitchen cupboard and easily see a collection of white mugs sitting against the white cupboard walls. The fact that these white mugs perfectly match their white background has no impact on our ability to see them clearly. The myriad tones and shadows provide all the information our eyes and minds need to make sense of the insides of our kitchen cupboards. Since light and dark contrasts are so pivotal in how we and other animals understand the world around us, it makes sense that animals have evolved ways of hijacking and overcoming these cues for camouflage.

Countershading:
camouflage in three dimensions

Even if a camouflaged animal perfectly matches the colour and pattern of their environment, predators can be very clever and have a few other ways to detect where animals are hiding. There is one animal feature that can be a dead giveaway, and it is something that every animal has – even you. It follows you around, everywhere you go. It's probably right behind you as you are reading this book. Do you know what it is? Your shadow.

There are two types of shadows that can give away an animal's location: cast shadows and form shadows. Cast shadows are the ones that follow you around. They are *cast* onto the ground by our bodies getting in the way between the sun and the earth. For a camouflaged animal, these can be a real pain in the neck. A superbly camouflaged bird might find their protective plumage completely useless on a sunny day when there is a suspiciously bird-shaped shadow on the ground next to them. The only way for animals to avoid casting a shadow when in full sun is to lie low. Camouflaged animals will press themselves flat against the ground, rocks, tree trunks or whatever surface they are standing on, to avoid casting a conspicuous shadow.

Form shadows are a little bit trickier to deal with. Form shadows are the shaded parts on a three-dimensional object that are on the side facing away from a light source. Hold your arm straight out in front of you. You probably have some light hitting your arm from above, maybe the sunlight, lamps, or overhead globes. The top side of your arm will be bright, while the underside of your arm will be shaded. As visual artists, Cott and Thayer would have understood this artistic principle; that to create the illusion of a three-dimensional object on a two-dimensional surface, you shade the underside of the object dark and leave the top light. In the natural world, this shading pattern of light on top and dark underneath would be a surefire clue to predators that there is a three-dimensional object there, no matter how closely the object matches the colour of its background. Cott and Thayer couldn't help but notice that animals have a clever way of countering this shading pattern called, funnily enough, countershading.

Countershading is a fancy word for when an animal is dark-coloured on top, and light-coloured underneath. Imagine a dolphin, for example: the top surface of a dolphin is often dark bluish-grey, much like the water around it, but the underside of the dolphin is bright white. You've probably noticed this in plenty of other animals, too. It's very common in fish and birds, so if you have a fish tank or an aviary at home, have a look and see if your pets are darker on top than they are underneath. Again, tigers are a perfect example of this; on their top side they are dark orange, on the underside they are bright white. There is a very good reason for this dark-to-light pattern: it counters the light-to-dark pattern characteristic of form shadows, resulting in a bit of optical trickery that hides the three-dimensionality of objects.

Look at the illustrations of a camouflaged fish on page 43. In the top row, it is evenly coloured and even though its grey colour matches the grey background, you can still tell there is something there because of the dark form shadow on the underside of the animal. No matter how closely that fish's colour matches its background, the telltale shadow can give away its position. Now, if that same fish were to have a bright white underside, the white counteracts and lightens the dark form shadow underneath, hiding the three-dimensional form of the fish. The overall impression is that the fish is now a flat uniform grey, just like its background.

Countershading is seen in all sorts of animals, across the entire animal kingdom. Palaeontologists have even shown that some dinosaurs were countershaded. Scientists have shown that fossils of a psittacosaurus and a feathered sinosauropteryx, both from China, had much darker pigments on their backs and lighter pigments on their undersides, consistent with countershading. (You might be wondering at this point, how on earth we know what colour a dinosaur was. Or any prehistoric animal, for that matter. We'll come back to that in a minute, let's just stay with countershading for a bit.)

Colour-changing cuttlefish will adjust their counter-shading depending on their orientation. Scientists experimented with sedated cuttlefish in the laboratory and found that when the cuttlefish were turned on their sides, or even upside down, their chromatophores would adjust so that the underside of the animal was lighter coloured and the topside of the animal was darker. Even under sedation, cuttlefish respond to their relative position, based on gravity, and adjust their countershading appropriately.

Even though the idea that countershading conceals form shadows has been around for over a century, scientists are still busily studying the finer details of how it works

Flat colour + Form shadow = Conspicuous

Countershading + Form shadow = Camouflaged

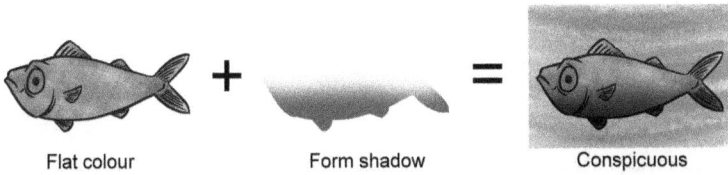

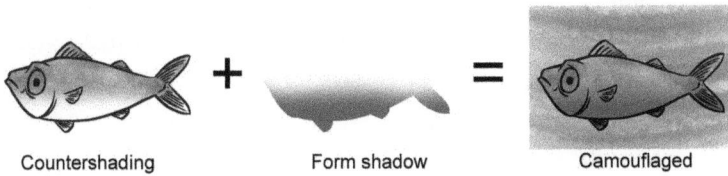

A quick guide to countershading

in the wild, with changing light conditions and different environments. When I was a student learning all about animal camouflage for the first time, I was taught a lesson about countershading that probably wasn't quite right: I was told that countershading exists because animals like fish and birds can be seen by predators from above *and* below. And if a predator were to look at a mottled brown bird from above, the bird would be camouflaged against the mottled brown dirt below. If another predator were looking at the bird from below, then the bird's bright white underbelly would blend in with the bright white clouds above.

I remember hearing this and thinking that something didn't quite make sense, and you might already be thinking the same thing. When I see birds fly overhead, they don't blend in with white clouds, they just look like dark silhouettes. Even something as brilliantly white as a seagull still looks like a dark black shape flying above. The same goes for fish. If you ever go swimming in the ocean and dive down underwater, you can look up at fish swimming above and see that they all look like dark black shadows against the bright sky above. There isn't much compelling evidence for countershading camouflaging animals against the sky, but you will still see the idea floating around books here and there.

However – there is always a *however*, isn't there? Or a *but* or an *exception to the rule* – in the animal world, things can always get that little bit weirder and more wonderful. There *are* animals that really *do* camouflage against the bright sky above. They don't do this using *countershading*, they do it using *counterillumination*.

But before we get to that, let's take a brief detour here to address that question of how we know what colour dinosaurs (or any prehistoric animals) were. Bones aren't the only structures that fossilise. Entire exoskeletons can fossilise in the case of prehistoric invertebrates, and sometimes skin and feathers can fossilise. These are the kinds of fossils that have started to change our perceptions of certain dinosaurs from scaly reptiles to feathery birds. In 2008, a team of scientists made an astonishing discovery when studying an unidentified bird fossil from the early Cretaceous: they found fossilised pigments within the feathers.

When you see a fossil in a museum, it's often displayed alongside a sculpture or painting of what we think the prehistoric animal would have looked like when it existed millions of years ago. When I go to a museum, I often stare at these two depictions, scratching my head as to how the scientists found their way from a sandy-coloured smush of bones embedded in a rocky slab, to the beautiful creature imagined next to it. It turns out that there is a plethora of information that can be mined from these bony smushes, if you know where to look. And where paleo-colour scientists look is very, very close. By examining fossilised feathers using a scanning electron microscope, scientists found that in certain areas, the surface was covered in small sausage-shaped lumps, each around 1–2 micrometres long. When they compared these microscale structures to modern-day bird feathers, the similarities were obvious. These tiny lumps looked identical to the pigment clusters found in modern-day bird feathers. They had just found fossilised melanin, which is the most common colour pigment in animals and gives colour to hair, skin, feathers, scales and eyes. And here it was, just as it exists today, bundled together and preserved in a fossil feather for around 54 million years.

Fossils with this level of colour detail are rare. Nevertheless, since then, fossil pigments have been found in dinosaur feathers and skin, prehistoric turtle shells, snake scales, ichthyosaur skin, insect exoskeletons and more. By sampling across entire specimens, scientists can show how pigmented patches occur in regular patterns, revealing the stripes, bands, spots and speckles of prehistoric animals. Not only can we tell where the pigments are, we can take a pretty good guess at what colours they produced. When melanin is packaged inside cells in elongated sausage-shaped bundles, it's associated with black or dark brown colour patches. When it's packed into small spheres, it's associated with reddy-yellow colour patches. There are a few cases, such as in a fossil of *Wulong bohaiensis* – a feathered dinosaur from the Early Cretaceous – where fossil imprints within the feathers show cylindrical bundles of melanin arranged in precise rows. These structures are usually associated with iridescent feathers so, as far as we can tell, about 120 million years ago there was a chicken-sized dinosaur flapping about the forests with glistening green-blue iridescent wings.

Up until now, paleo-artists were given a great deal of creative licence to imagine what dinosaurs and other prehistoric animals would have looked like. But now that we can piece together fragments of information from fossils with preserved pigments, we are beginning to appreciate these extinct animals for the beautifully and intricately coloured creatures that they were. While we can never truly know what these animals looked like and how they behaved, every piece of new information we get brings our blurry view of the prehistoric past into slightly sharper focus.

Counterillumination: glowing down for looking up

As we learned earlier on in this book, as you travel into the depths of the ocean, the weirder the world starts to become. Prepare yourself for some more weirdness as we travel once more into the darkness of the deep. We won't travel too far down this time, just to around 500 metres, to a part of the ocean we call the twilight zone. Above us, near the water's surface, there is plenty of sunlight making everything visible. Below us, in the deep ocean there is no sunlight, and all creatures must live in complete darkness. Here, in the twilight zone, only small amounts of blue light penetrate this deep, so the whole world consists of dark black below and dim blue above.

If a sea creature living at this depth looks up, they won't see the water's surface or the sky above them, they will just see a faint blue glow. If they look down, they will see the dark black of the deep ocean. This means that to camouflage in the twilight zone, an animal might appear dark black (or red, as we saw in the previous chapter) to match the colour of the ocean abyss below. This will protect them from predators that look down on them from above. But when you're floating in the middle of the ocean, predators could be watching from any angle. If a predator looks at their prey from below, they could still see them easily as a dark silhouette against the dim blue glow from above. How do deep-sea animals blend in against blue glowing light from above? It's simple: they glow blue, too.

In the darkness of the deep ocean, there are numerous strange creatures that produce bioluminescence. Many species of fish, squid and crustaceans that live in the twilight zone have bioluminescent organs that cover their undersides. They glow a dim blue colour that matches the blue light that shines down from above. This means that any predators looking at the animal from below won't see a dark black silhouette. The animal bodies essentially become glowing projection systems that help them hide in the glow of the twilight zone. Their bioluminescence comes from special organs called photophores. Inside these organs are special cells where chemical reactions occur that give off light. Some animals have small clustered photophores, like deep-sea shrimp that have patches of glowing cells under their heads and abdomens. Some deep-sea sharks and squid have undersides that are entirely covered in tiny photophores. Up close, the photophores look like an array of blue dots, but from a distance, their glow merges together to form a solid blue light, like pixels on a screen.

The brightness of the blue light from above varies with depth. The deeper you go, the fainter the light is, until eventually all light disappears, and even dim bioluminescence will stand out in the darkness. Counterilluminating animals have clever ways of switching on and off their bioluminescence and can adjust their own brightness, like a dimmer switch on a light bulb. Lantern shark's bellies are covered in thousands of tiny photophores that can be switched on and off using hormones. When the light from the lantern shark's belly is turned off, the photophores are covered in a layer of dark pigment. When it's time to turn its lights on, the lantern shark releases hormones that cause the pigment layer to open, revealing a glowing blue dot in the centre of each photophore. The pigment layer of each photophore now looks like a dark black ring, and the more these rings expand, the brighter the lantern shark's belly glows.

To adjust their bioluminescence, animals need to measure the light coming from above using special light-sensitive organs. The photophores of the deep-sea shrimp *Janicella spinicauda* seem to be able to both emit light *and* sense the ambient light levels. Some deep-sea lantern fish have evolved a strange solution to this by having a photophore positioned on their head that they can use as a reference, like a painter comparing colour swatches. These lantern fish have bellies covered in photophore spots and, weirdly, a single photophore inside their head. This glowing photophore sits inside the fish's skull and points directly towards the eye, so it is never actually seen by other animals. The light from the photophore passes through a special transparent gap around the eyeball before passing through the eye lens. Scientists' best guess is that these fish are using their eyes to look at the ambient light above and comparing it with the glowing photophore inside their heads. They then modify the intensity of that photophore to match the light above, and the photophores on their undersides follow suit.

During the day, counterillumination only works in the twilight zone where the light coming from above is very dim. As impressive as bioluminescence is, the light it produces is very faint. Above the twilight zone, the dim glow of bioluminescence won't counter the brilliant light of the sun. But at nighttime, there are animals that use counterillumination as a camouflage strategy up near the water's surface, where the dim glow of their undersides can help them blend in against the glowing night sky.

During the day, Hawaiian bobtail squid (*Euprymna scolopes*) hide by burying themselves in sand. At night, they come out of hiding to hunt. Just like other animals use counterillumination to camouflage in the twilight zone, these squid have special organs on their underside that glow, matching the overhead illumination. But, unlike the examples given above, these squid don't produce bioluminescence using chemical reactions. In fact, the squid don't produce any light at all. Their bioluminescence comes from colonies of glowing bacteria that live inside their organs.

Remember how I said to prepare yourself for more weirdness? Well, here we go. Just like other counter-illuminating animals, these bobtail squid can adjust the brightness of their glowing bacterial organs, and even switch them on and off depending on light conditions. They can do this because the organs that house their glowing symbiotes aren't just simple sacs full of bacteria. For counterillumination to work, the squid need to have complex adjustable organs. Surrounding a chamber of glowing bacteria there is a layer of shiny, mirror-like tissue that stops any light from escaping upwards by reflecting it all downwards. This light then must pass through a thick layer of transparent tissue that filters and focuses the light. You have probably heard that squid have ink sacs that shoot out thick dark ink. These bobtail squid have modified ink sacs that surround their bacterial light organs. By injecting ink into channels that run around the light organs, they cover areas of the glowing bacteria with ink, cloaking the light like the iris of an eye. Unlike eyes which absorb light, this special organ has evolved to emit light. And at the end of each nighttime hunting session, when the squid don't need their glowing bacteria anymore, they just squirt them out. Using special ducts in their light organs, the squid empty out almost all the bacteria. Then, while they are hiding under the sand during the day, they grow an entirely new colony of glowing bacteria in preparation for the next evening's hunting session. So, for things like nocturnal squid, counterillumination isn't just as simple as having a glowing blue patch on your belly. It means having a rechargeable biological projection system complete with reflectors, filters, lenses and diaphragms, powered by symbiotic bioluminescent bacteria. Weird, huh?

Camouflage in complex environments

Up until now we have been talking about animal camouflage in relatively simple settings, and I have been asking you to imagine how an animal of a uniform colour, maybe with a bit of countershading, can disappear against a background of a similarly uniform colour. Unless you happen to be reading this book while hiking through the vast snowfields of Siberia, or the flat desert plains of central Australia, you'll notice that the natural world around you isn't just a flat colour swatch for animals to blend in with. Look around forests, seagrass beds, coral reefs, rocky outcrops or any other environment you can imagine, and you will see that it is a complex collage of innumerable colours and tones. In the understorey of a forest, bright leaves and branches contrast against dark trunks and soil. Light filtered through the canopy litters every surface with speckles of light. The surface of a single rock could appear glaringly bright where it is hit by direct sunlight, and almost black where it lies in shadow.

To camouflage in complex environments, animals have evolved their own complex body patterns. Take wildcats for example – scientists have shown that species like leopards and jaguars, which have complex spots and rosettes, are more likely to rest in the forest canopy, where their complex coats may help them blend in with the complex background of leaves, branches and speckled lighting. Species with plain coats, like lions and cougars, are more likely to be found in the open. Animals display all kinds of complex spots, stripes, bands, speckles and flecks that come in a huge range of varying patterns with different levels of complexity. The idea isn't necessarily to have a coat that matches the features of your background, but to have a complex pattern that draws attention away from the body outline.

Let's use fish as an example again. Imagine a fish like the ones in the top half of the illustration on page 54. A uniformly coloured fish swimming through a complex seagrass bed is quite conspicuous. Even when the fish's tone matches parts of its background, the telltale shape of a fish is clear to any animal on the lookout for fishy-shaped things. Once we imagine fish having more complex patterning, the outline of its body is much harder to distinguish. This is still true even when the pattern of the fish isn't a particularly good match for the background.

This phenomenon is called *disruptive colouration*. It works by creating contrasts within the animal's body that are greater than the contrast between the animal and its background. These usually involve patterns that run perpendicular to the upper and lower edges of the animal, creating false edges that distract from the outline of its body. The end result is that the animal no longer appears to have its telltale shape, and it can be easily overlooked as its contrasting patterns become jumbled up with the contrasting patterns of the background.

Scientists from the UK conducted an ingenious study using artificial moths, which were essentially small brown paper triangles with black patches on them. These were placed outside on oak trees and the scientists monitored what types of disruptive patterns were more or less likely to be attacked by birds. Artificial moths with black patches that intersected their outlines were better camouflaged than the ones where the black patches didn't interrupt their outline.

Disruptive colouration is so effective it appears to work even when the colours don't match the background. The peach blossom moth (*Thyatira batis*) is brown with conspicuous pink spots on the edges of its wings. In a similar study to the one above, scientists created artificial moths that had pink and brown disruptive patterns or more

Disruptive colouration in action

plain brown patterns. When placed outside on complex mossy backgrounds, the disruptively patterned pink and brown moths were less likely to be attacked by birds than the plain brown ones. They took this one step further and created completely artificial-looking blue and pink moths, and even these completely unnatural-looking moths were less likely to be attacked than the plain brown moths.

New Zealand lichen moths (*Ipana atronivea*) have zig-zagging black stripes that contrast against bright white patches. They virtually disappear when placed against lichen, and scientists have shown that they are much less likely to be attacked by predators when on lichen than on other surfaces. The scientists also found that from the perspective of a predatory bird, the black and white patches of the moth weren't a particularly good colour match to the colours of lichen. Instead, they found that the complexity of their patterns was a good match to the complexity of the patterns naturally formed by lichen, and this may be more important for survival than being an exact colour match to the background.

Again, we can understand how and why disruptive colouration works using art as an analogy. Painters and illustrators know that our eyes are drawn towards areas of high contrast. In *Girl with a Pearl Earring*, Vermeer contrasts the almost white skin of the girl's face against the dark black background, immediately drawing our attention towards her face. Meanwhile the more subtle contrasts between the background and her golden shawl or blue headdress are details that we only notice once our eyes start to wander around the painting, looking for more detail. It is this bias towards areas of high contrast that disruptive colouration seems to exploit, as it causes eyes and minds to overlook the outline separating an animal from its background.

This isn't to say that colour contrast never comes into play, but we must remember that we aren't the best judges when it comes to how colour works in nature. Since we have excellent colour vision, we often assume that colour differences are equally important to other animals. But, as we covered in the previous chapter, many other animals, including most mammals, aren't that great at telling apart different colours. Monochromats, which have only one type of photoreceptor, see the world in monochrome so brightness differences may be the only visual information they can use to distinguish objects in their environment.

Disruptive markings don't just include complex spotty and stripy patterns. Big, bold patterns can also work in similar ways. The giant panda is perhaps the most famous example of this. Their black legs and shoulders contrast strikingly against their bright white back, neck and head, making them look distinctly different to any other closely related bear species. By analysing photographs of giant pandas taken in the wild, scientists have shown that their dark and light patches blend in with different parts of their forest environment. The black bands blend in well with shadowy patches, and the white bands with bright snow or illuminated surfaces. Thus, the panda's patterns serve to break up their body into different components and conceal their bear-like shape from natural predators such as tigers, leopards and dholes (Asiatic wild dogs). If this works for pandas, it's easy to imagine how the same might be true for other boldly patterned animals like tapirs, colobus monkeys or African okapi.

Not all stripes are equal

Bold stripes and patterns don't just seem to draw attention away from animals' outlines. Bold lines often seem to be placed in ways that could draw attention from other conspicuous bodily features that, if spotted, would be a dead giveaway to predatory animals in the environment. Many animals have dark stripes that run directly across their eyes and faces. It is thought that the dark spots of animal eyes are such noticeable features that these dark facial stripes have evolved to make eyes less conspicuous. The disruptive eye masks of raccoons are perhaps the most famous, but the more you look at animals the more you start to notice how common it is for them to have dark stripes running across their eyes. They can be seen in birds, insects, reptiles, amphibians, fish and more. Other stripes appear to distract from the form of limbs and tails.

Most of the research into why animals have contrasting patches, stripes and spots comes from studies like the ones mentioned above, on camouflaged moths or moth-like triangles cut out of paper. Some of the more famous examples of boldly patterned animals are large mammals like cheetahs, tigers, tapirs, badgers and orcas. These are, for obvious reasons, much harder to study in the laboratory than moths and pieces of paper, and in many cases we still don't have direct evidence for why these large animals have such interesting patterning. The idea that disruptive eye masks, for example, can help an animal camouflage is still arguably an untested hypothesis. There are other possible explanations for why animals have eye stripes and masks. These dark patches may reduce the amount of glare reflecting off their faces, helping the animals see in bright sunlight. In 2009, Professor Tim Caro from the University of California, Davis, wrote a paper about how little we know about the conspicuous patterns we find on large mammals and concluded that the research had 'barely moved beyond anecdotal stages of investigation'.* In many cases this is still true, and even some of the most iconic spotted and striped animals have presented real challenges for scientists trying to figure out why they look the way they do.

If I asked you to name the first striped animal you could think of, what would you say? I would put good money on it being one of two things: a zebra or a tiger. These conspicuously striped animals, living on opposite sides of the planet, are some of the most iconic and easily recognised animals in the entire animal kingdom. Look in any board book for young babies and, nine times out of ten, you'll find a picture of a zebra or a tiger as an example of something that has stripes. It's no surprise that these two animals have been the focus of much discussion around how stripes function in nature. However, they have also proved to be some of the most mysterious examples of why animals have stripes.

---

*   Tim Caro (2009) Contrasting colouration in terrestrial mammals. *Philosophical Transactions of the Royal Society of London* 364, p. 537.

Zebra stripes seem to break all the rules we just established. Unlike the examples above, zebras don't live in complex environments like dense woodlands, they live in open grassy plains. So, their black and white stripes aren't a form of disruptive colouration, like the black and white patches of a giant panda. When we adopt the perspective of a zebra's natural predators, their stripes seem even stranger. Lions and hyenas don't have vision as sharp as humans do. If you were to look at a zebra from, say, 100 metres away, the black and white stripes would still be perfectly visible. Lions and hyenas have much less acute vision, so from the same distance the stripes of a zebra would start to blur together into a vague grey mass. Add to this the fact that lions and hyenas tend to hunt at dusk and dawn, when low light levels would make zebra stripes even harder to make out.

People have been trying to make sense of zebra stripes for centuries and have come up with a long list of possible reasons. Perhaps their stripes act like fingerprints and let zebras recognise individuals within a herd. Maybe stripes confuse or dazzle a pursuing predator while the zebra is in motion – this isn't as crazy an idea as it might sound; we'll deal with this dazzling idea later. Scientists have even examined whether the white and black stripes would absorb different amounts of heat in the sunlight and create little swirling currents of air at the surface of a zebra's skin, helping to keep them cool in the heat of the African savannas. These ideas have some merit, but there hasn't been any particularly convincing evidence for any of them. But there is mounting evidence for another somewhat strange hypothesis about the function of zebra stripes. It is becoming clear that stripes *do* protect zebras from bloodthirsty predators: not lions or hyenas, but bloodsucking tsetse flies.

Tsetse flies are native to tropical Africa and feed on the blood of large mammals like horses, cattle, pigs and sometimes humans. Using their long and pointed mouthparts, they pierce an animals' skin, tap into a blood vessel and drink the free-flowing blood. Their bite, however, while painful, may be the least concern for any person or animal bitten by a tsetse. These flies can transmit numerous bloodborne diseases including, in humans, the sleeping sickness known as trypanosomiasis. Tsetse flies are a major concern for livestock producers, as animal trypanosomiasis is estimated to kill somewhere around 3 million cattle every year in Africa. Compared to their domesticated relatives, zebras aren't commonly attacked by tsetse flies in the wild and it seems to have something to do with their stripy coats. Tsetse flies, along with other common biting flies like horse flies, stable flies and horn flies, are much less likely to land on striped animal coats than on uniformly coloured coats. These flies seem to prefer animals with dark-coloured coats over light-coloured coats and are seemingly repulsed by striped coats. Scientists still aren't 100 per cent certain how this works, but we know that it does. Experiments have shown that striped zebra pelts left outside have fewer flies land on them than on solid-coloured animal pelts. Even inanimate objects painted with black and white stripes have much lower numbers of flies landing on them than plain black or white objects. Scientists in Japan showed that painting black cattle with white stripes can reduce the number of biting flies that land on them.

We know that flying insects often use contrasting edges as important landmarks when navigating or co-ordinating their flight paths, so it's possible that zebras' bold black and white stripes play tricks with flies' eyes and confuse or overwhelm their senses as they fly past. Another possibility, which has been put forward by a team of scientists in Hungary, is that the stripes could confuse flies' thermal detection behaviour. When a parasitic fly lands on their host, they need to find a blood vessel underneath the skin to feed from. To do this, they may use heat-sensing organs to detect slight variations in temperature, where the skin above the blood vessel is slightly warmer than the surrounding skin. Because the black zebra stripes absorb more sunlight than the white stripes, they are slightly warmer. This turns the surface of a zebra into a confusing network of criss-crossing heat gradients. As a result, the flies might simply find it too hard to accurately detect blood vessels to feed on, and habitually avoid striped zebras as unrewarding feeding grounds. While evidence overwhelmingly points towards zebra stripes having an anti-parasite function, rather than an anti-predator function, the nuts and bolts of how this works are still being figured out.

And finally, on to tiger stripes. Do they help tigers disappear into the forest through disruptive colouration? Potentially, yes. The take-home message about tiger stripes is: we don't know for sure. Tigers are a bit of an anomaly. Out of all the large predatory cats, only tigers have bold vertical stripes. The rest are plain coloured (e.g. lions), spotted (e.g. leopards), or have irregular patterning (e.g. clouded leopard). There is some evidence that the size and spacing of tiger stripes generally resemble the complexity of their jungle background. And given our understanding of how disruptive patterning works, it seems intuitive that they must help conceal a tiger's outline. But to date, no one has been able to provide any rigorous proof. The fact that the iconic stripes of this iconic species are still a bit of a mystery stands testament to how little we sometimes know about animal camouflage and how it works in the wild.

Animal camouflage was a focal topic in the work of pioneering biology rockstars like Darwin, and the theoretical principles of animal camouflage were put forward by Cott and Thayer over a century ago. Even though these principles have been studied and discussed for generations, there is still so much to learn. That great chasm of the unknown grows unimaginably large when we start to consider how camouflage works in senses beyond our own. So far, this book has focused primarily on camouflage in the sense of visual perception. Most of the research literature shares that same bias. We are, after all, visual creatures, so it is easiest for us to make observations and ask questions about things that we perceive with our eyes.

Though I have made some references to how camouflage can involve wavelengths of light beyond those that we can see, like infrared and ultraviolet, camouflage can work in other sensory realms like scent, sound and electromagnetism. It is relatively easy for us to comprehend how things like background matching and disruptive colouration can trick our visual systems, but it becomes much harder to imagine how similar processes could be happening chemically or acoustically. And it becomes near impossible for us to imagine how camouflage works when it pertains to senses that we don't even have the organs to detect, like electromagnetism or thermal signatures. When it comes to understanding animal camouflage beyond the world of light and vision, scientists are only scratching the surface of what is happening and how it works.

Camouflage in all senses

We don't have to look very far to find an example of an animal whose sensory experience of the world is strikingly different to our own. Dogs, our closest animal companions, are famous for their incredibly sensitive noses. It's worth remembering that while we share our homes and lives with dogs, they are living in a completely different sensory universe. They can't see the range of colours that humans do, can't distinguish differences in brightness as well we can, and their vision is nowhere near as sharp as ours, but their sense of smell beggars belief.

A recent study showed that dogs could be trained to detect the scent of eucalyptus when diluted down to a ridiculous concentration of around one part per one hundred quintillion. When following a person's scent, tracker dogs can figure out the direction of travel based on the difference in scent concentration across as little as five footsteps. They can be trained to sniff out hospital patients that have different types of bacterial infections, and forms of cancers, diabetes and oncoming epileptic seizures.

Despite this impressive resume, even a dog's keen sense of smell can be tricked by chemical camouflage. Scientists from South Africa trained dogs to search out six different species of snakes purely by scent. For five of these snake species the dogs could sniff them out with over 80 per cent accuracy, but one species left the dogs completely stumped. African puff adders (*Bitis arietans*) are ambush predators that look amazingly camouflaged hiding in leaf litter. They are also apparently chemically camouflaged, as the trained sniffer dogs seemed completely unable to detect them by scent.

A similar experiment was conducted with meerkats that live in the same habitat as African puff adders. They were equally stumped by their chemical camouflage. How this works still isn't clear. It's possible that puff adders' low metabolism means that the scents they give off are in small amounts and very faint in odour.

Other cases of chemical camouflage may work through some form of chemical background matching. The coral-feeding filefish *Oxymonacanthus longirostris* takes refuge within the branches of the corals it feeds on. As they feed, the fish absorb the corals' chemicals and somehow incorporate them into their tissues, giving them the scent of corals. This makes the filefish harder to detect by predatory cod. Similarly, the larvae of tortoise beetles (*Chelymorpha* spp.) take on the chemical profile of the plants they are feeding on, protecting them from predatory ants.

It's likely that chemical camouflage is widespread in animals. It's also likely that it works in complex ways that our noses simply don't comprehend. While I'm sure you're great at appreciating the nuanced tones of your favourite bottle of plonk, if I asked you what a snake smelled like, I doubt you would have a great answer. The chances then of humans noticing that some snakes smell a bit less snakey than others are probably very slim.

While we are woeful at navigating our world through scent, humans are much better at understanding the complexities of sound and how animal calls might be modified for auditory camouflage. Many social animals rely heavily on sound for communication with their families and social groups. This comes with the inherent risk of attracting the attention of eavesdropping predators, and there is growing evidence that many animal calls may be adapted to avoid detection from predators. Several different dolphin and porpoise species communicate using a limited range of high-frequency clicks that are beyond the hearing abilities of killer whales. Similarly, when southern right whale mothers are rearing newborn calves, they modify their calls to be relatively quiet and infrequent, presumably to reduce the likelihood of attracting nearby killer whales to their vulnerable young. Asian corn borer moths (*Ostrinia furnacalis*) may use a similar trick, by calling to females with short-range ultrasonic calls that are less conspicuous to hunting bats.

Graeme Ruxton, a modern-day biology rockstar (the Pink Floyd of animal camouflage research, if you will) has been at the forefront of drawing attention to camouflage beyond the realm of vision and the difficulties we face in understanding how it works. Since our framing and understanding of animal camouflage is firmly rooted in the science of animal vision, we might not even have the language to describe camouflage in other senses. Consider the example of Blainville's beaked whales, which also appear to modify their calls in response to the threat of killer whales. These whales dive deep into the ocean together as a social group. When they are deep underwater, and out of the range of killer whales, they call to each other in a chorus of clicks, rasps and buzzing sounds. When they return to the surface and are potentially within earshot of killer whales, they go silent. While this appears like other examples of auditory camouflage, Ruxton argues that silent animals don't count as an example of auditory camouflage. Instead, it is more analogous to an animal hiding under a rock than to an animal that is visually camouflaged. Sound energy behaves very differently to light energy, so applying the same concepts to both visual and acoustic camouflage quickly becomes problematic.

Beyond sight, sound and smell, we struggle to even imagine what it must be for animals that rely on completely different sensory stimuli. Platypus can catch prey in complete darkness by sensing the electromagnetic fields emitted by small animals buried underground. Snakes can pinpoint the location of prey using thermal heat signatures. Some seals can use their whiskers to follow the trails of swirling water left behind by swimming fish.

Understanding the basics of how these bizarre senses work is fundamentally challenging. Thus, taking a further conceptual leap into imagining how such senses could be tricked and sidestepped is a form of mental gymnastics currently beyond our comprehension. Throughout the rest of the book, I will challenge you to think further outside your own sensory abilities and we'll explore the many tricks that nature uses to fool eyes, noses, ears, antennae and other sensory organs.

From here on in, we extend far beyond the realms of animal camouflage into other forms of deception. Because while some animals get by fine simply hiding from predators, others take a much bolder approach and have evolved elaborate disguises that conceal and confuse.

3

Mimics and masquerade: hidden creatures on parade

You might be surprised to hear that scientists are still furiously tinkering away in their laboratories and on their computers, trying to understand how animal camouflage works. It seems like something we should have figured out by now, doesn't it? Camouflage is never as simple as it seems, and new discoveries about animal trickery are being made all the time. An entirely new type of animal camouflage was demonstrated as recently as 2010, when a team of scientists from the UK demonstrated a clever type of animal trickery called 'masquerade'.

Lots of the animals we have been talking about so far have a rather vague similarity to their environment. A sandy-coloured crab hiding on a sandy seafloor can be tricky to spot, but if you took that same crab and put it on some green leaves, its cryptic camouflage doesn't work anymore – now it just looks like a sandy-coloured crab. Some animals have such sophisticated camouflage that they can avoid this problem. They don't just have a vague resemblance to their backgrounds, they are such impressive tricksters that their entire bodies take on the form of a specific part of their environment. Stick insects are perhaps the most famous example. They have long thin bodies with thin legs that make them look just like small twigs. This makes them near impossible to spot when hidden among other twigs, and, unlike other cryptically camouflaged animals, if you take them out of this habitat and put them somewhere else, they still look like twigs. The Indian oakleaf butterfly (*Kallima inachus*) is famous for its eerie similarity to a dead leaf. Its wings come to narrow points at the front and back in perfect affinity with the shape of a tapered leaf. Their mottled brown colours with a dark stripe that runs between the front and back wing points creates a superb illusion of a dry flat leaf. Many grasshoppers and leaf insects look almost identical to bright green leaves. The veins of their wings look just like the veins of a lush thick leaf, and some even have ruffled and curled edges to their wings that make them look as though they have been chewed on by a caterpillar. There are frogs that look like rocks, lizards that look like dead leaves, fish that look like seaweed, and even spiders and caterpillars that look like wet glistening bird droppings.

Decades ago, scientists interested in animal camouflage pondered whether those animals with extra impressive camouflage tricks were somehow different to the rest and described them as having a 'special resemblance' to their environment, whereas other cryptic animals had a more 'general resemblance' to their environment. Now, we call this special resemblance 'masquerade'. Even though we could describe, from a human perspective, the difference between crypsis and masquerade, we didn't know if or how it worked in nature. Maybe we were the only animals silly enough to think that stick insects looked like sticks? Or if stick insects did get ignored by predators was it because the predators didn't see them in the first place, or did the predator see the conspicuous-looking insect but decided that it was a stick? This raised all sorts of philosophical questions about how animal brains worked. Do animals even know what sticks are? When a predator looks around their environment, do they simply categorise things as food or not-food, or do they have a complex understanding of their environment with different mental categories for things like sticks, leaves, rocks and soil? If they do, do they have to learn these things or are they hardwired into their brains from birth?

These questions went mostly unanswered until 2010, when Dr John Skelhorn and his team devised a clever experiment that showed, for the first time, how masquerade really worked. In this experiment, they studied camouflage in twig caterpillars. The larvae of brimstone moths (*Opisthograptis luteolata*) and early thorn moths (*Selenia dentaria*) look just like short stumpy twigs, complete with rough-textured skin in shades of mottled green and brown. They sit very still and rest with their whole body extended from a tree branch, making them look like little stumpy twigs. To find out what predators thought about these twig caterpillars, Dr Skelhorn and his team decided to ask the most ferocious predator they could get their hands on: baby chickens.

Using chicks that had just hatched, they could assume that they had no experience with the world and then ask them what they thought of both twigs and twig caterpillars.

They released the chicks into a plain white walled enclosure and presented them with twigs. At first, they would start pecking at the twigs as if the twigs were food. Eventually the chicks learned that they weren't food and stopped trying to eat them. Then, when the chicks were given a live twig caterpillar, they didn't try to eat it. Even though they could see the caterpillar clearly in the plain white enclosure, they didn't recognise it as food. The chicks had learned what twigs were, and after they had done that, they mistook the twig caterpillars for twigs and wouldn't eat them.

This means that masquerade works in a completely different way to other forms of camouflage. Masquerade doesn't work by avoiding *detection*, it works by avoiding *recognition*. A masquerading animal can be out in the open and spotted easily by predators but can still be safe because their disguise tricks the predator's brain into thinking that they are something totally different. Since this pioneering study, the same methods have been adopted to demonstrate how masquerade works as a camouflage strategy in other animals like dead leaf butterflies (*Kallina* spp.) and bird-dropping spiders (*Phrynarachne ceylonica*).

Another species of twig caterpillar, *Biston robustum*, has been shown to take on the chemical profile of their food plant, which protects them from predatory ants. This raises the possibility that masqueraders might not just *look* like their environmental counterparts, but also *smell* like them. Another intriguing possibility is that masquerading animals could evolve to *behave* like their models. In many cases, this is straightforward. A twig-like caterpillar can behave like a twig by staying very still in a twiggy kind of way. But what happens if the thing that you are supposed to be masquerading as starts to move? Anyone who has ever held a stick insect has probably noticed that when they do move (which admittedly isn't very often) they make curious side to side motions with their whole body. For a long time, people have thought that this might be a special behaviour to help them blend in among leaves and sticks swaying in the wind. A team of scientists from Australia tested this in the Australian spiny leaf insect (*Extatosoma tiaratum*, also known as Macleay's spectre), and showed that wind currents triggered the insect's swaying motions. They also moved more when wind speeds were variable rather than constant.

An even more dynamic example of camouflaged animals seeming to behave like parts of their environment comes from the tropical rainforests of South-East Asia. Here you can find gliding lizards (*Draco cornutus*) that leap from high in the canopy and soar long distances. While resting on a tree trunk, they blend in with the mottled grey-brown trunk surface. When it's time to fly, they push themselves off the trunk and open wide the wings that seem to emerge miraculously from either side of their bodies. These wings are formed by specialised elongated ribs that fan out and stretch the skin between them into a fine membrane. Seen from below, the sunlight shines through the thin skin of their membranous wings and they seem to glow in bright shades of red, orange and yellow.

Australian scientist Danielle Klomp led a team of scientists studying draco lizards in Borneo and found that the bright colour of their wings matched the colour of dry leaves in the local environment. Lizards with rusty-red wings tended to live in areas where there were lots of

Mimics and masquerade: hidden creatures on parade

trees that shed rusty-red dry leaves. Other lizards with dark green to black wings lived in areas where trees shed dark green dry leaves. This led the scientists to suggest that perhaps the lizards are mimicking falling leaves as they fly, allowing them to stay camouflaged while on the move by masquerading as a falling leaf in motion.

Masquerading animals are often referred to as 'mimicking' particular objects, be they leaf mimics, twig mimics or whatever the case may be. The word 'mimicry', though, usually refers to a different type of deception in nature, where sometimes the trick is not to hide away and look uninteresting, but to stick out like a sore thumb.

Yellow and black says stay back

As a rule, camouflage requires stillness (except, clearly, for the above examples where it doesn't). Even the most perfectly camouflaged leaf insect starts to look suspicious once that leaf appears to sprout legs and walk away. For highly mobile animals, blending in isn't an option and other defensive strategies are required. These could be defensive strategies, including hard shells, agile speed, sharp spines or tough scales. Or slightly more offensive tactics, like sharp teeth, potent venom and poisonous chemicals. The best weapons, though, are the ones that you never have to use, which is why many animals advertise their defensive arsenal with bright conspicuous colours. We only need to wander through a garden to see these in action.

The contrasting bright yellow and black stripes of bees and wasps are perfect examples of an adaptation called

73

aposematism. It's a mouthful of a word and not particularly intuitive, so I'll refrain from using it and instead stick with the more intuitive term *warning colours*. Any predator that messes with a venomous bee soon learns to associate yellow and black bands with a nasty sting and should learn to stay away from other bees in the future. Bees' conspicuous stripes are an unambiguous warning to any would-be predator. Other classic examples include the vivid colours of poison arrow frogs, the telltale spots of distasteful ladybugs and the neon hues of toxic sea slugs.

Colours aren't the only signals that animals use to warn predators of their toxicity. The green and black poison frog *Dendrobates auratus* also has a telltale scent that predators like snakes learn to avoid. Tiger moth caterpillars sequester toxins from the plants that they eat, and incorporate the toxins into their adult moth body tissues. Their toxicity is advertised by bright warning colours on their wings and abdomens. This, however, does little to protect them from predatory bats which hunt in the darkness of night. Bats locate prey in complete darkness using sonar – they make high-frequency soundwaves and listen for the echoes that bounce off solid objects. To protect against echo-locating bats, tiger moths produce their own high-frequency sounds that act as a kind of acoustic aposematism, or warning call. They have special sound-producing organs that emit loud clicks that ward off predatory bats.

If you go looking for animals with warning colours, you start to notice some common themes. Yellow and black stripes are a good example; many bees, wasps, hornets, caterpillars, bugs and grasshoppers use a similar signal to tell predators to back off. Bright orange or red contrasted against black is another common combination used by many chemically defended animals. This makes sense for two reasons: first, bright warm hues against dark black make for a very easy to read signal. Subtlety is of no use here. Demure tones and gentle gradients don't make very effective warning signals. It's probably no coincidence that the warning signs humans make, from street signs to product labels, use similar colour palettes. Secondly, it makes sense that an animal would evolve warning colours that already exist in the environment, rather than develop entirely new ones. Any predator that has learned to avoid the yellow and black bands of a bee will most likely apply that same reasoning in avoiding other animals with similar colour patterns.

A warning colour pattern shared between multiple species is reinforced in the minds of predators encountering many different prey species, so natural selection should result in animals converging upon a shared warning colour system. This phenomenon was first suggested by Fritz Müller in 1879 when describing two distantly related butterfly species, both of which were chemically defended against predators and had converged upon the same warning colour pattern. We now call this phenomenon of shared warning colours 'Müllerian mimicry', and the cohort of species that share warning colour pattern are often referred to as being part of 'mimicry rings'. The best-studied example of mimicry rings are *Heliconius* butterflies in South America. Here, dozens of different butterfly species showcase their warning colours to predators, and the patterns they use depend on the geographical region and habitat. In some areas, different species have converged upon yellow stripes on a black background, whereas in other habitats they display black spots on bright orange, or bright red-orange patches against black.

Now, wait a minute – I know what you're thinking: this book is supposed to be about deception and trickery. This is all lovey-dovey kumbaya nonsense. Animals honestly signalling their trickery to predators, giving them a heads-up so no one gets hurt. Plus a bunch of animals sensibly sharing a defence system and mutually benefiting from each other's presence. You're right – this isn't deceptive at all. But Müller wasn't the first one to notice similarities in warning colour patterns between different animals. Before Müller ever conceived of what we now call Müllerian mimicry, another man set the stage for understanding what mimicry is and how it works. He would become the namesake for a more deceptive form of mimicry. His name was Henry Walter Bates.

Batesian mimicry

Perhaps it's fitting we're comparing evolutionary biologists to classic rockstars. Henry Walter Bates and his travelling companion Alfred Russel Wallace are among the most legendary of biology rockstars. Charles Darwin came from a wealthy and educated family and his voyage on the HMS *Beagle* was sponsored by the British government. Bates and Wallace, on the other hand, were a pair of Leicestershire lads with a yearning for adventure. They pulled themselves up by their bootstraps and funded their own globetrotting research expeditions.

In their twenties, in 1848, they hopped on a merchant ship to the Brazilian Amazon, where they got down to business, collecting and cataloguing thousands of species. They kept a portion of the specimens they collected for themselves and for science, and the rest they sold to British natural history museums to continue funding their work. Bates would persist and explored the remote Amazon rainforest for a whopping eleven years. Wallace left after four years, taking his collection of specimens with him on another merchant ship headed for England. Three weeks into the journey, the ship caught fire and the crew, including Wallace, abandoned ship, along with his collection. They spent ten days adrift at sea before they were rescued by another passing merchant ship. Now *that* is mettle.

Around this time, people broadly accepted how natural selection could result in animals camouflaging within their environment. What hadn't yet been sufficiently explained was why some other animals didn't seem to do this at all. What could explain the gaudy feathers of birds and vivid wings of butterflies? In the past, many people appeared to genuinely believe that the colours of animals and plants were there purely for people's enjoyment. Wallace put it wryly in his 1877 paper entitled 'The Colours of Animals and Plants' when he asked: 'What could be the use to the butterfly of its gaily-painted wings, or to the humming bird of its jewelled breast, except to add the final touches to a world-picture, calculated at once to please and to refine mankind?'* He would continue in this paper to thoroughly

---

*   A.R. Wallace (1877) The colours of animals and plants. *Macmillan's Magazine* 36, p. 384.

throttle that suggestion and explain in detail how animal and plant colours were products of natural selection. Wallace was the first to use the term 'warning colouration' and explain how these vivid colours and patterns acted as a threat to predators.*

This idea was likely in Bates's mind as he trudged through the Amazonian jungle, collecting insects. He found many tropical butterflies that had foul-tasting chemicals in their tissues. Some were from the Danainae family, often called milkweed butterflies, because they feed on toxic milkweed plants and incorporate the plant toxins into their own chemical defences. He watched as birds would either avoid eating these butterflies or attempt to eat them before quickly dropping them, due to their foul taste. Milkweed butterflies often had bright conspicuous yellow and black colour patterns advertising their toxicity. He also found butterflies from a completely different family, the Pieridae, that looked almost identical to the milkweed butterflies. It took Bates's skilled eye as an entomologist to tell the two types apart. From a distance, the pierids looked to be just another milkweed butterfly. It was only by catching them and examining them up close that he could identify them as a completely different species.

This intrigued Bates for several reasons. Unlike the milkweed butterflies, these pierids didn't have foul-tasting chemicals inside that would explain their bright warning colours. Furthermore, in other parts of the world,

---

*   Aposematism isn't the only explanation for bright colours in animals. Sexual selection, for example, can lead to bright colours used in mating displays.

pierid butterflies had more humble colour patterns, often uniform white or yellow. Here in the Amazon, the pierids had bold bright colours with dark black bands that bore a strange resemblance to a completely different type of butterfly. Bates realised that if the two butterflies looked similar from his perspective, then they probably looked similar in the eyes of other animals. He assumed that if birds had learned to avoid toxic milkweed butterflies, they would also cautiously avoid the similar-looking pierids.

Here was the explanation for their similarity. Since warning colours alert predators to an animal's toxins or other defences, another animal that mimics these colours gains the same protection, except their display is a bluff. They don't have the bite to match their bark. Bates called the similarities between these two animals 'mimetic analogies', and he termed the undefended Pieridae butterflies 'mimics' and the distasteful Danainae butterflies their 'models'. These terms are still used over a century later. We'll meet other forms of mimicry later in the book, but this form of mimicry, where an undefended 'mimic' resembles a defended 'model', is named in Bates's honour as *Batesian mimicry*.

Upon reading Bates's published work, Charles Darwin wrote a letter to Bates, lauding his work and announcing it as 'one of the most remarkable and admirable papers I ever read in my life'.* Bates's discovery filled in a noticeably missing puzzle piece in the grand theory of evolution via

---

*   Charles Darwin, Letter to HW Bates (20 November 1862), in
     F Darwin (ed.) (1887) *The Life and Letters of Charles Darwin*,
     vol. 2, John Murray, p. 76.

natural selection. In many ways, Bates's description of deceptive mimicry set the foundation for most of the ideas you will read in this book. It established that deception and trickery could form the basis of survival strategies beyond simple camouflage.

The discovery and acceptance of Batesian mimicry opened the floodgates for naturalists looking to explain and describe similarities between two entirely different species. Numerous other butterflies have since been shown to mimic other chemically defended species, but Batesian mimicry extends far beyond butterflies and spans the animal kingdom. Hoverflies (*Syrphidae* spp.) are a classic example. These common garden flies are easy to spot as they wear the bright yellow-and-black disguise of bees and wasps, yet are completely harmless. Research has shown that birds will avoid attacking hoverflies if they have previous experience with bees. The North American red-backed salamander (*Plethodon cinereus*) varies between speckled orange and black and bright uniform orange, and they appear to be Batesian mimics of a co-occurring toxic salamander which, as juveniles, have bright orange aposematic colouration.

One of the most famous and widely discussed examples of Batesian mimicry is in venomous coral snakes and their non-venomous mimics. (So broadly discussed, in fact, that it's become a contentious issue and Batesian mimicry might not be the only reason for their conspicuous banding patterns. But more on that in the next chapter.) Dozens of venomous coral snake species have brightly coloured banding patterns with varying combinations of alternating red, black, yellow and white stripes. These are believed to function as a warning colouration, advertising their highly potent venom which can be fatal to large mammals, including humans. Other non-venomous snakes have similarly bright banded patterns but don't necessarily match coral snakes in the order and combination of colours that make up their patterns. Nevertheless, the bright colours are enough to be suggestive of a warning signal and so they are considered Batesian mimics of coral snakes. This has led to much discussion and testing of how important it is for putative mimics to precisely match the features of their models or whether 'close enough is good enough' to fool potential predators.

When humans are trying to distinguish coral snakes from non-venomous lookalikes, differences in their colour banding sequence can sometimes be a helpful indicator. Discussions of coral snake mimicry often call upon a rhyme that is apparently used in North America to help distinguish venomous snakes from their mimetic counterparts:

Red on yellow, kill a fellow.
Red on black, venom lack.

While it's supposed to be memorable jingle to help people know whether they've encountered a venomous snake, it's not exactly the catchiest earworm ever written, is it? What was it again? Red on yellow, happy fellow? Black on red, you'll be dead? Yellow on black, good for Jack? I dunno. Besides, the mnemonic only refers to a handful of non-venomous species in the northern United States that have red stripes that touch black stripes (or was it yellow?)

and not the dozens of other coral snakes and their mimics across the globe that have different banding patterns. Your best bet is to let Batesian mimicry work its magic and just don't go near big brightly coloured snakes. Leave that one to the professionals.

Mimicking a highly venomous snake seems like a good strategy for an animal to use to avoid unwanted attention. If you had to pick a dangerous animal that would be the perfect thing to mimic, what would you choose? Something that strikes fear into the hearts of other animals that, if you were mistaken for one, would surely guarantee your protection. I know what you're thinking: those venomous creepy crawlies that skitter through the undergrowth. With their long legs, empty black eyes and telltale body shape, they are immediately recognisable as fearsome predators. They are, of course, ants. (What!? You were going to say spiders? No! Our spider fears are mostly cultural inventions that aren't shared in the opinions of other animals. Don't believe me? I've written a whole book about it; you should check it out.* In fact, spiders are so benign that there may even be an animal that uses spider mimicry to *lure in* other animals, as opposed to *repelling* them. Keep reading, we'll get to that story.)

As far as the rest of the animal kingdom seems concerned, ants are some of the scariest creatures around. They bite, sting, spray acid and attack in thousand-strong swarms. Ants are so scary that all manner of animals have

---

* There are plenty of fascinating spider stories and facts in the book, *Silk & Venom: The incredible lives of spiders*, such as flying spiders and adorably fluffy jumping spiders, and what spiders do when they dream.

evolved bizarrely modified body forms to look more like ants for protection. Ant mimicry is common in jumping spiders, whereby they develop constricted abdomens and elongated thoraxes that resemble the segmented shape of an ant's body. Often the spiders will wave their first pair of legs in front of them as they walk, in an affectation of wriggling insect antennae. Many insects, particularly as immature nymphs, look markedly different from their adult form and, instead, look just like brightly coloured ants. Numerous praying mantises, stick insects and true bugs make near-perfect imitations of formidable ants. Even some ants appear to be Batesian mimics of other more toxic ants. In Malaysia, a relatively harmless *Camponotus* ant species shares the contrasting black and yellow patterning of the chemically defended ant *Crematogaster inflata*. Birds avoid eating the *Camponotus* ants if they have previously encountered the distasteful *Crematogaster* ants.

In Australia, over a hundred different ant species have conspicuously coloured iridescent gold abdomens. This appears to be a warning colouration, and other dangerous insect species, such as a number of stinging wasp species, have converged upon the same iridescent gold abdomen warning colour. Scientists have described this as a Müllerian mimicry ring, where the shiny gold colours are reinforced as a signal to ward off predators. Other animals seem to have jumped on the golden-butt bandwagon as Batesian mimics of golden ants. Seven harmless species of spider, three bug species and one plant hopper species have all been found to make convincing mimics of golden ants. Together these animals create a large mimicry complex that begins to blur the lines between Batesian mimicry,

Müllerian mimicry and when warning colours may or may not be honest signals of an animal's defences.

Camouflage and mimicry aren't just for animals

When it comes to being eaten, plants are sitting ducks. Immobile and rooted to the ground, plants are supremely vulnerable to any herbivore that slowly meanders past, munching on leaves and stems. When we talk about camouflage, we have an inherent bias where we assume that plants are innocuous things that animals hide among, but there is no reason why plants shouldn't have evolved their own deceptive strategies for survival.

Lithops are a type of succulent often called the 'flowering stone'. These minute plants are round, mottled grey and look remarkably like stones. When they don't have a gaudily bright flower poking out of their middles, they blend in beautifully with the rocky arid soils that they grow in. Some cactus species such as *Sclerocactus papyracanthus* are coated with a layer of what looks like tufts of dry grass, which may disguise the cactus as dead and withering vegetation.

To acknowledge that plants are vulnerable to being eaten is not to say that they are defenceless. Scores of plants fill their tissues with poisonous toxins and cover their surfaces with razor-sharp spines and thorns. Some advertise their defences with bright colours, using the same warning colour strategy as bees, snakes and frogs. Many cacti and spiny succulents (e.g. *Agave* spp.) have brightly coloured spines. Thorny bushes such as roses, brambles and briar plants have conspicuously coloured thorns that may serve to advertise their weaponry to curious herbivores.

Just like in animals, there are harmless plants that have seemingly evolved to mimic dangerous ones for protection. A survey of plants in Israel found several species with either sharp spines and thorns, or toxic chemicals, which also showed contrasting white and green variegations (e.g. milk thistle, *Silybum marianum*) suggestive of a mimicry ring where these different species use similar signals to showcase their defences. The survey also found four seemingly undefended plants (*Lannaea* spp.) that have similar bold markings on their leaves, and appear to be Batesian mimics of the harmful plants.

Some plants (e.g. *Xanthium strumarium* and *Arisarium vulgare*) can develop conspicuous dark spots that have all the hallmarks of a bold visual display that says 'back off.' These plants aren't toxic, and don't appear to mimic any other harmful plants in the environment. Simcha Lev-Yadun, an Israeli botanist and the mind behind many ideas in defensive plant mimicry, believes that a very different type of mimicry seems to be happening here. He suggests that the mottled black spots could create the illusion of a plant crawling with ants. There are many plants that form a tight mutualistic relationship with ants, whereby the plant provides food and shelter for the ants in return for the ants acting as in-house security. They crawl all over the plant surface and feed on herbivores like caterpillars and small bugs. It's not too big a stretch of the imagination to think that another plant could protect itself by adopting the disguise of an ant-defended plant.

We should stop for a second to consider who these mysterious herbivores are that are supposed to be tricked by these illusions. Small insects like caterpillars are a real danger to plants so having a strategy to repel them would make sense. However, caterpillars don't have great eyesight and probably aren't gazing across a dense forest, picking which plant they are going to slowly crawl towards next. Caterpillar mothers, on the other hand, might be choosier about what plants they pick, and there is some evidence to suggest that these defensive tricks might work by scaring off butterflies looking for a good place to lay their eggs.

When butterflies are picking a spot to lay, the optimal spot should be the one that gives their offspring the best chance of survival; that is, a leafy plant that provides lots of food for their baby caterpillars when they hatch. Obviously, a plant that looks like it's crawling with ants wouldn't be suitable, and neither would one that looks like it's already got a bunch of caterpillars on it. If a plant can create these illusions, it may have a way to deter butterflies from laying small armies of leaf-eating caterpillars on its surfaces. Several species of passion vines (*Passiflora* spp.) develop conspicuous yellow spots either on their leaf tips or across the leaf surface. They look suspiciously like insect eggs, and greenhouse experiments showed that butterflies are less likely to lay their eggs on or near leaves that have these yellow spots. This has been described as a form of 'egg mimicry' that works to deter egg-laying butterflies.

The South American plant *Caladium steudneriifolium* has conspicuous white lines that meander across the surface of its bright green leaves. These lines look somewhat like the meandering lines left on leaves by an infestation of mining moth larvae and could serve to deter other egg-laying moths. Scientists tested whether these white marks might mimic a moth infestation by painting white lines on otherwise healthy-looking green leaves. When they returned to inspect the leaves a few months later, the modified leaves were much less likely to have moth larvae on them than the plain green leaves.

Perhaps the most outlandish example of what could be a butterfly deterrent signal is seen in the semaphore plant *Codariocalyx motorius*. As we've already established, plants don't move, except for the ones that do. And the semaphore plant is, like all other plants, immobile, until it's not. The small paddle-shaped leaves of this shrub tend to move in a way that leaves don't. They bend at the stem and rotate, drawing slow circles in the air. I say slowly, but as far as plants go, the semaphore plant sets a blistering pace. At top speed the leaves can complete a full rotation in about 90 seconds. A possible explanation for this (which also comes from the fertile mind of Simcha Lev-Yadun) is that the slowly turning leaves give the illusion of butterfly wings dappling the surface of these small shrubs. The idea is that any butterflies looking for a place to lay their eggs would see this bush, think that it is already covered in butterflies, and fly elsewhere looking for less competitive territory to lay their eggs.

I've not seen a semaphore plant in action, save on a few YouTube clips. And at the risk of sounding like an ignorant lurker in the comments section, I'm not convinced. The butterfly wing illusion takes quite a bit of imagination to perceive. There may well be an alternative explanation. Perhaps the leaf movement makes it hard for the butterflies to alight on the leaves, and the deterrent is physical rather than visual. Then again, I am not a butterfly and don't see the world through butterfly eyes. Plus, when it comes to making the next great discovery in camouflage and mimicry, an open and imaginative mind is a must.

I'll ask you now to embrace your imaginative minds. We are about to go on a journey into some recent research that is throwing scientists a few intellectual curveballs. It has changed the way we think about deception in plants, and perhaps the way we understand the nature of plant life altogether. If you aren't seated in a comfy chair already, perhaps it's time to find one. If you have a roll of tinfoil nearby, feel free to sculpt it into a rudimentary hat and put it on. Defensive mimicry in plants is a field of science with a lot of exciting ideas but a paucity of good data. We're on shaky ground here when it comes to cold hard facts, and the discoveries described below have been met with equal parts excitement and incredulity. They have scientists stumbling over their words as they try to describe plants that sound like they are straight out of science fiction.

No one expected the Patagonian vine conundrum

Mistletoes are ectoparasitic plants that grow on the trunks and branches of other larger trees. In Australia, there are mistletoe species that grow on things like eucalyptus, acacia and casuarina trees, where they absorb water and nutrients from the tissues of their host plants. There are a few dozen Australian mistletoe species whose leaves seem to strangely match the appearance of their host plants. Mistletoe species that grow on eucalyptus trees tend to have elongated teardrop-shaped leaves like their host trees, whereas species that grow on casuarina trees have long pine-needle-like leaves, just like casuarina. The same goes for mistletoes that have adopted the thin pointed leaves of their acacia hosts. Scientists pondering why this might be the case have pointed vaguely to some sort of undefinable transfer of 'morphogens' between the two plants, as if the mistletoes were sucking the essence from their hosts and taking on their forms.

In the 1970s, Australian scientists Brian Barlow and Delbert Weins put forward a different idea that could explain these apparent similarities. It was no less contentious than the morphogens idea and persists to this day. They suggested that the mistletoes were mimicking their host plants, to avoid being eaten by herbivores. Unlike the cases above, where plants disguise themselves as something harmful or distasteful to avoid being eaten, the idea that a leafy green plant would mimic another leafy green plant for protection seems a little harder to wrap your head around.

The leaves of trees like eucalypts are tough, waxy and oily, and are generally hard for animals to digest. Unless they are a highly specialised eucalyptus feeder, like a koala, herbivorous animals would be better off looking for softer, more palatable green leaves. Mistletoe leaves, it was believed, must be comparatively edible, so it's possible that looking like a hard waxy eucalyptus leaf makes them seem very unattractive to herbivores. But what herbivores? Animals like koalas, and many insects, are perfectly happy to eat eucalyptus leaves, so mimicry wouldn't offer any protection against those herbivores. Australia isn't home to any tall, grazing herbivores other than kangaroos, and they mostly eat grass.

Having eliminated most other candidates, Barlow and Weins deduced that possums must be the ones falling for mistletoe mimicry. They even floated the idea that perhaps mistletoe mimicry had evolved to fool animals that don't exist anymore. Australia was once home to megafauna – enormous marsupials that could have foraged on leaves high up on trees. They point to examples such as *Procoptodon* and *Simosthenurus* – giant kangaroos that stood over 2 metres tall. Perhaps they were responsible for mistletoes evolving to look like their host plants, and the mimetic patterns we see now are the lingering shadows of our prehistoric past. It's a beautiful thought, but a completely untestable hypothesis now that these animals are extinct. Despite not having any concrete evidence to back up their hypothesis, there was no better explanation for mistletoe mimicry, and the idea was cautiously accepted.

Since then, follow-up studies haven't offered much support for the idea. Field studies found that mistletoes suffered similar levels of herbivory regardless of whether they mimicked their host plants or not. Some have questioned whether possums eat mistletoes in the first place. Meanwhile, attempts to quantify shape and colour similarities have failed to find any measurable similarity between mistletoes and their hosts, which raises the genuine possibility that their apparent similarities are happy coincidences or figments of our imaginations.

Since its initial proposal, the intriguing idea of parasitic mistletoes mimicking their host plants has fallen out of favour. That was until recently when a new discovery injected some vigour back into this old idea. Not only that, it's conjured up a suite of botanical possibilities that many scientists had placed either in the too hard basket or the silly ideas pile.

In 2014, Ernesto Gianoli and Fernando Carrasco-Urra published a report on a climbing vine from southern Chile. What they described had never been seen before, nor since, and scientists have embraced the discovery like poking a sleeping bear with a stick – giddy with excitement but keeping a safe distance from what might turn out to be a crazy idea. *Boquila trifoliolata* is a long, woody vine that climbs and curls around other trees. Gianoli and Carrasco-Urra reported that where the vine weaves through the leaves of other trees, the vine leaves grow to match the shape, colour, size and venation of their supporting tree. Just like mistletoes, the vine leaves seem to mimic their supporting tree's leaf shapes. It's slightly more complex though, because where the vine isn't in close contact with another plant, the leaves take on a more 'normal-looking' vine leaf shape. That idea on its own is pretty interesting, but the story doesn't end there.

Unlike mistletoes, these vines are not ectoparasites and they don't have a specific relationship with a particular host plant species. Like many crawling vines, they climb and weave themselves around anything nearby. Which means that *B. trifoliolata* leaves don't just grow to resemble one type of leaf, they can grow to mimic a whole range of different leaf types, from the small round succulent leaves of prickly myrtle bushes to the long dark leaves of tall ulmo trees and the pointy serrated leaves of mitre flower bushes.

What's more is that a single vine draped across multiple trees will have different types of leaves along its length depending on what types of plant it encounters.

But (with the danger of sounding like a cheesy television infomercial) wait, there's more! Not only do the *Boquila* vines mimic leaves of other native plants in their habitat, but they have also been seen to mimic the leaves of introduced plants like the creeping buttercup (*Ranunculus repens*), which is only known to have been introduced to the area a few decades ago – not long enough to have influenced the evolutionary trajectory of the mimetic vine. In which case these vines are more akin to a colour-changing chameleon or octopus than the Batesian mimics described earlier, albeit changing form on a much slower timescale. This unprecedented discovery has earned *B. trifoliolata* the common name of chameleon vine.

As incredible as these findings are, the most important part of this research occurred when Gianoli and Carrasco-Urra measured rates of herbivory on different leaves. By measuring how much had been eaten of the different types of leaves, they could assess which were more appealing to leaf-eating animals: the 'normal-looking' vine leaves or the disguised mimetic leaves. The pattern they found was clear: the non-mimetic vine leaves were much more susceptible to being eaten than the disguised leaves. Here was the first evidence that mimicking the leaves of another plant in the vicinity could protect a plant from being eaten. The phenomenon first hypothesised for Australian mistletoes (and never fully supported) was now found to be plausible in a Patagonian climbing vine.

With a likely explanation for *why* climbing vines mimic

other plants, attention then turned to the more confusing question of *how* they mimic other plants. Whereas mistletoes are physiologically intertwined with their host plants and draw nutrients from their host's tissues, climbing vines merely use other plants for physical support. They curl around and drape over surfaces, which means that some kind of direct biological transfer of information through roots or other tissues is unlikely. Furthermore, the scientists found that a physical connection to a plant wasn't even necessary for mimicry to occur. Where vines climbed around the bare trunks of trees, they still had non-mimetic leaves. Mimicry only took place where the vine was near leaves. And those leaves didn't need to belong to the plant supporting the vine. In parts where a neighbouring tree had overhanging leaves that came close to the vine, the vine seemed to mimic the leaves of the neighbouring tree. Close proximity to leaves seems to be the simple trigger for mimicry.

With this in mind, Gianoli and Carrasco-Urra put forward two bold explanations for how these climbing vines were detecting and responding to nearby leaves: plant volatiles and/or horizontal gene transfer. The first idea is relatively straightforward. They suggested that the nearby leaves may be giving off some kind of airborne chemicals that the *Boquila* vines detect, and that information then triggers certain biological processes that lead to their own leaves taking on the appearance of nearby leaves. The second idea of horizontal gene transfer is quite a bit more complex. To avoid lengthy tangents into hardcore genetics and wading dangerously into complex sciences I don't grasp very well myself, I'm going to gloss over some details here.

Put simply, horizontal gene transfer is a crazy ability that things like bacteria and viruses have. Certain bacteria can take genetic information from other organisms, like other bacteria that they have just attacked, and place it neatly into their own genome. This is how bacteria are capable of rapid evolution of new traits like antibiotic resistance.

What does this have to do with a leaf-mimicking vine? There could be some system whereby bacteria living in or on nearby plants incorporate certain parts of those plants' DNA into their bacterial genome. Then, should those bacteria become airborne, or come into contact with the vine, the genetic information from the nearby plant gives the vine the information that it needs to morph its leaves into the correct mimetic shape. It sounds bananas, but in 2021, Gianoli and a team of colleagues published a report showing that the bacterial communities on the surfaces of mimetic leaves were highly similar to the bacterial communities on the nearby leaves they were mimicking, and distinct from the bacterial communities on non-mimetic leaves. This suggests that the vines' mimetic leaves share a microbiome with their nearby model plants, giving some credence to the possibility of horizontal gene transfer occurring between the two species.

Just as the tantalising bargains of a bombastic info-mercial keep coming, so do the twists and turns of this story. One other slightly bonkers idea that has arisen from this discovery was that the chameleon vine might somehow be able to *see* the leaves nearby and use this information to determine their leaf shape. Plant vision is an idea that was bandied about and then thrown out over a century ago. There are certain types of algae that have specialised cells that can respond to light energy, but no one has ever discovered any dedicated light-sensing organ, the equivalent of a 'plant eye', hidden somewhere in the tissues of plants. The incredible behaviour of the chameleon vine has inspired some scientists to ask whether we should look again.

In 2022, scientists Jacob White and Felipe Yamashita figured that since we're wading through the realms of ludicrous ideas, why not throw another one into the mix. To test for the possibility that *B. trifoliolata* can see nearby plants, they thought, why not grow these vines in a greenhouse with plastic artificial plants and see what they do. Incredibly, the chameleon vines modified their leaf shapes in places where the vine overlapped with the artificial plants. They didn't quite match the artificial plant's leaves in size and shape, but there was a clear pattern where the leaves diverged from their usual vine-leaf shapes. Here, in a glasshouse experiment, devoid of any shared bacterial communities or airborne plant chemicals, the chameleon vines were still able to somehow detect and respond accordingly to the presence of nearby (albeit fake) leaves.

To summarise the whole story, we have no idea what's going on. If these plants can see, there is no currently accepted idea for how this is possible. Though the idea that plants could detect and respond to light information is not that crazy. Their whole existence revolves around utilising light to gather energy. Plants turn their leaves to face the sun, they direct resources so that they grow towards light sources. When you think about it, plants are covered in dense networks of leaves and photosynthetic surfaces whose entire purpose is to capture light energy. The idea that this light energy could constitute information that is processed to make sense of their external environment is, in a sense, the definition of vision.

Perhaps a vivid imagination is required to make the next great leap in scientific understanding. After all, it took the vivid imagination of Henry Walter Bates to conceive of what Charles Darwin himself couldn't. Just like it took the creative minds of Cott and Thayer to approach animal camouflage from a different angle and propose some bold ideas which were to become broadly accepted concepts. All it takes is for an outside-the-box idea to be supported by a whole lot of good evidence and suddenly everyone agrees that they always thought it was a perfectly reasonable idea in the first place.

Since Bates and Wallace's pioneering explorations of how natural selection could lead plants and animals to evolve survival strategies that rely on deception beyond simple camouflage, many more strange and curious forms of natural trickery have been described and hypothesised. So far, we have talked about using trickery to avoid predators altogether. But these strategies aren't fool proof. Camouflaged animals eventually get spotted, and predators are sometimes willing to risk tasting that brightly coloured, potentially poisonous prey. What happens once that predator decides to move in for the kill? That's a whole new ballgame where new kinds of trickery come into play!

There is a spider in this photo, I promise. The combination of background colour matching, mottled patterning and a textured body outline pressed up flat against the trunk surface makes this lichen spider (*Pandercetes* sp.) near invisible, even when it's right in front of your nose. *James O'Hanlon*

This Australian sea lion (*Neophoca cinerea*) demonstrates how countershading may help obscure the three-dimensional shape of an animal. The delineation between the grey colouration on top sits right along where the form shadow lies on the underside of the animal. This results in both the top and bottom of the animal having a similar grey tone. *James O'Hanlon*

We can get a rudimentary idea of how some animals see the world by modifying photographs. Since digital images are made of red, green and blue colour information, we can take this digital image of a tiger and remove the information from the red channel to get an idea of how an orange tiger might be trickier to spot for an animal like a deer that has only red and green photoreceptors.

Modified from image by Wikimedia user:
Kalyanvarma - Wikicommons CC BY-SA 4.0

The lichen moth (*Ipana atronivea*) has disruptive black bands that make it difficult to see the outline of the moth against the complex lichen background. *Cassandra Mark Chan*

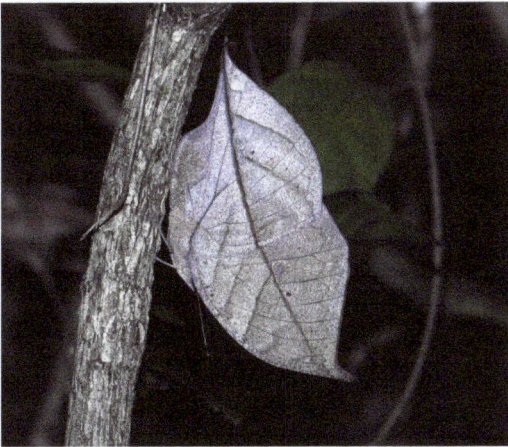

The dead leaf butterfly (*Kallima inachus*) is famous for its unbelievable likeness to a dry leaf. *Dr. Raju Kasambe – Wikicommons CC BY-SA 4.0*

Caterpillars of the citrus swallowtail butterfly (*Papilio aegus*) start off looking like wet glistening bird droppings and eventually grow into a more cryptic green colouration. *James O'Hanlon*

The Australian spiny leaf insect (*Extatosoma tiaratum*) sways back and forth when stimulated by a gust of wind. It doesn't just look like its background, it behaves like it, too. *James O'Hanlon*

On the top row are two remarkable ant mimics: an Alydid bug (left) and a jumping spider *Myrmarachne macleayana* (right), and on the bottom row are the actual ant models: (left) the green tree ant *Oecophylla smaragdina* and (right) *Polyrhachis robsoni*. *Jim McClean*

The mountain katydid (*Acripeza reticulata*) lifts up its wings to reveal startling blue and red colours to would-be predators. *Michael Whitehead*

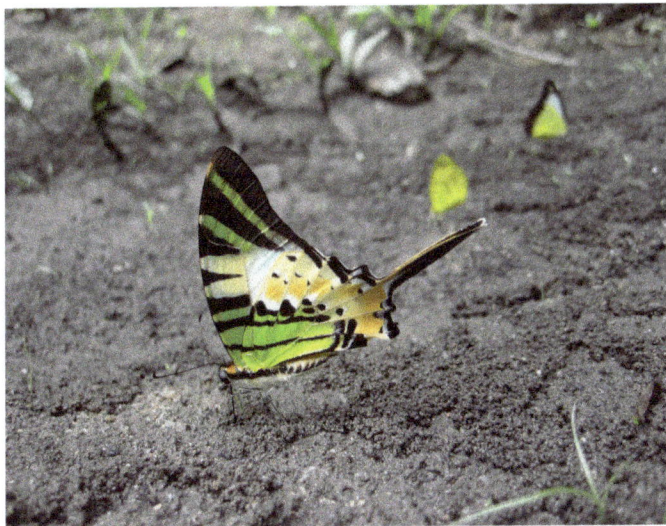

This five-bar swordtail (*Graphium antiphates*) is facing left, but the illusion of false antennae on the right, formed by extensions of its wings, is believed to direct predator attacks away from its head. *James O'Hanlon*

This threadfin butterflyfish (*Chaetodon auriga*) shows two of the adaptations thought to misdirect predator attacks away from the head: a dark stripe concealing the fish's real eye, and a false eyespot on the rear end of the fish's dorsal fin. *James O'Hanlon*

A dead leaf praying mantis (*Deroplatys* sp.) showing off the startle display that it uses to ward off predators. *James O'Hanlon*

A flower-like orchid mantis (*Hymenopus coronatus*) feeding on an unlucky honey bee. *James O'Hanlon*

The orchid mantis's floral disguise attracts pollinators as prey for the mantis. *James O'Hanlon*

Cryptostylus tongue orchids release scents that lure in male *Lissopimpla excelsa* wasps who attempt to mate with the flower. In this image you can see the yellow pollinia that the male picks up attached to the tip of his abdomen. *Amy Brunton-Martin*

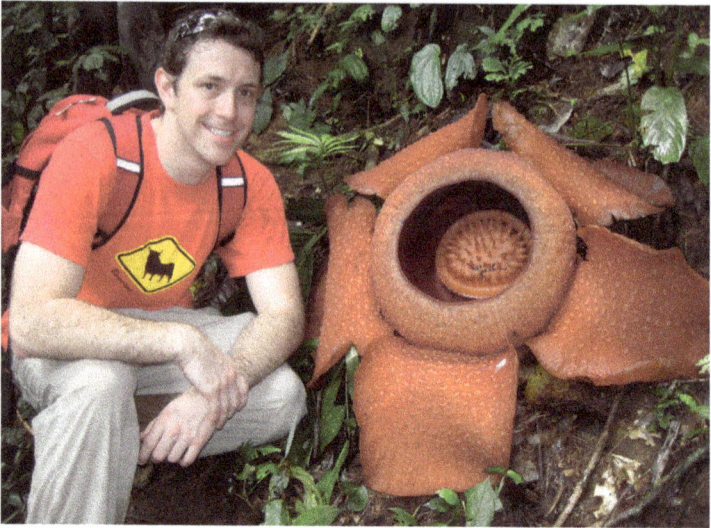

Rafflesia flowers are the largest single flowers in the world and draw in insect pollinators by smelling like rotting carcasses. I had the pleasure of seeing this majestic flower in the Cameron Highlands in Malaysia. *James O'Hanlon*

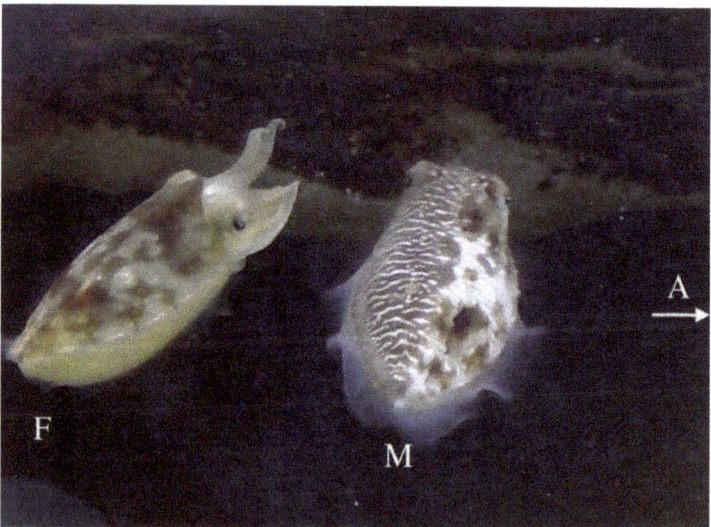

Male mourning cuttlefish (*Ascarosepion plangon*) flexibly change their colour patterns to deceive rival males. The courting male (M) shows his striped mating display to a female (F) while simultaneously disguising himself as a female to an adjacent male (A). *Martin Garwood*

4

The art and magic
of misdirection

Of all the forms of deception that are covered in this book, the one that we should be most familiar with is misdirection. And if you aren't familiar with it, then, well, that's precisely the point. Misdirection happens when our attention is drawn away from what really matters to something more innocuous. When done well, we should be completely oblivious to the fact that it has occurred. Think about every time you have watched a politician on television making a grand big hoo-ha about a particular topic of their choosing, while savvily avoiding other pressing and much more important issues. Think about your last trip to the supermarket, when you arrived with a short list of ingredients for a healthy recipe you found online; perhaps some tahini sauce, an eggplant and whatever the hell nutritional yeast flakes are. Then, the next thing you remember is walking out of the grocery store with a receipt as long as your arm and a trolley full of chocolates, ice cream and nary a yeast flake in sight. And then, when you get home your kid spots the bags full of ice cream and the only way you can distract them is with the flashing colours and sounds of a cartoon on TV.

We think we can trust our senses, we think we are masters of our own behaviours and the decisions we make, but our attention is constantly pulled in all directions. From the piercing 'ding' of our smartphone notifications to the perfect symmetry of some golden arches on signs dotted along main roads – these are all examples of the ways that control of our behaviours can be subtly taken away from us. The simplest way to do this is to use the in-built biases of our senses to pull and push our attention in different directions.

A good example of this might be, say, you're reading a book. And the small text clustered into paragraphs is what you're here to read. It's where all the juicy stories and information lie. But, rather annoyingly, there are clusters of big bold text that draw your eye away from what you are supposed to be reading. You know that you're not supposed to read it. It says so right there in big letters. You know that you are supposed to read top to bottom, left to right, but you couldn't help yourself. You probably read the big bold words first and worked your way past to the small lower contrast ones before finally getting to this paragraph here. It's not your fault, it's just what our eyes do. Big bold shapes with high contrast contours instinctively draw our gaze much more than finer scale shapes with low contrast.

DON'T READ THIS LINE LAST.

ALSO, DON'T READ THIS.

Stage magicians have advanced the craft of misdirection into a complex and flamboyant artform. Innocuous gestures like a wave of a hand, or a turn of the head, can cause spectators to look towards the shiny top hat on the table, and not the plain handkerchief being quietly stuffed into the magician's back pocket. Entire crowds of people can be fooled into believing that everything from a shiny coin to a live elephant can magically appear on stage from out of nowhere. By taking masterful control over what people do and don't pay attention to, magicians make impossible things seem possible right in front of our scrutinising eyes. These same games of misdirection are played in nature. Just like big bold text can draw your eyes away from a good story, animals can use bold contrasting patterns to play games with a predator's attention.

DON'T READ THIS!

Detachable deflective tips

If there is anything we have learned from watching scores of comic book superhero movies, it's that if you want to take down the bad guy, you should go for the head.* It seems that predatory animals know this as well and will aim their attacks towards the vulnerable head of their prey. Just like big contrasting blocks of text can draw your attention away from the story you're supposed to be reading, animals can use bold colours to draw attention away from their heads. And the further away from the head the better.

Many lizards have conspicuously coloured tails that deflect predator attacks away from their heads and bodies, towards their tail tips. The most famous are perhaps the blue-tailed skinks (genus *Plestiodon*) that look like someone has taken a common garden skink and dipped its tail into a pot of vivid blue paint. Their incongruous-looking tails effectively deflect predator attacks towards their tails. As you probably know from chasing skinks around the garden as a kid, lizard tails are expendable. In most species, they will readily fall off and eventually grow back. Sometimes lizards will drop their tails pre-emptively, leaving their

---

* As in Marvel's *Avengers: Infinity War*, where an injured Thanos gloats to Thor and mutters menacingly, 'You should have gone for the head', before unleashing the power of the infinity gauntlet with a click of his fingers. I don't really need to explain this movie reference, do I? It's one of the highest grossing movies of all time – chances are you've seen it already. But for the small fraction of you that haven't seen the film (which you should, it's really good!), here's an explanation for you. Speaking of misdirection and distraction, what on earth are you doing still reading this long and completely unnecessary footnote – get back to the book!

squirming and brightly coloured tails as a lure for the predator while they scurry away to safety.

Many butterflies use the same trick. Swallowtail and swordtail butterflies have long, conspicuous filaments that extend outwards from their hind wings. Like the true tails of lizards, these false 'tails' on butterfly wings can be conspicuously coloured and are easily detached without having a great impact on the butterfly's ability to fly. Just like their reptilian counterparts, these butterflies' 'tails' deflect predator attacks away from their vulnerable bodies at the relatively minor expense of some damaged wing tips.

Certain moths, such as the luna moth (*Actias luna*), have long elaborate 'tails' on their wings, similar to butterfly wing tips. This might seem strange at first, given that moths are nocturnal. A conspicuous false target seems like it would serve little purpose under the darkness of night. That is, until you remember that moth predators don't hunt using their eyes, they hunt using their ears, listening for the sonar echoes of fluttering moth wings.

A team of American scientists did an experiment on luna moths to see whether their long wing tips helped them avoid capture by bats. When moths had their 'tails' fully intact, bats were only able to catch them around 34 per cent of the time. When the researchers removed the moths' tails the capture rate shot up to 81 per cent. The moths' wing tips were creating echoes that made it harder for the bats to hit their target. When they watched high-speed footage of the bats catching the moths in mid-air, the scientists saw that the bats were more likely to aim their attacks towards the tails, rather than the moths' bodies. A follow-up study dived deeper into what kind of echoes the wing tips produced. They found that the echoes the tail tips produced were relatively quiet compared to the rest of their wings. So, they weren't overly conspicuous, like the bright tail ends of butterflies and lizards. It's almost as if the moths dragged behind them a small and awkward footnote to their echoes, a bit like a large unnecessary footnote on the page of a book. It's something not quite obnoxious enough to overwhelm the main message but niggling enough to throw the bats off their game and ever so slightly distract them away from the main body.

Ahem, my eyes are up here …

Earlier on, we talked about how animals often have bold stripes that run across their heads and seem to conceal their eyes. In principle, this should help camouflage the animal since big round glossy eyes are a dead giveaway no matter how well camouflaged the rest of the animal's body may appear. Many butterfly fish elaborate on this tactic by not only having bold stripes that cover their eyes but also having large conspicuous false eyes down near their tails. Lab studies using artificial prey found that predatory fish direct their attacks towards eyespots, rather than eyes concealed with a black stripe. Many butterflies (the flappy flying ones, as opposed to butterfly fishes) have large circular eyespots on the margins of their wings. Like the conspicuous wing tips of other butterflies, these eyespots can deflect predator attacks towards the wing margins.

The combination of conspicuous false eyespots, and a relatively inconspicuous head region, sometimes results in a backwards-looking animal, with a head where its tail should be and a tail where its head should be. If the illusion works, not only will a predator be tricked into attacking a less vulnerable part of their prey, but will be a little extra confused if their prey darts off in the opposite direction to the way they were supposed to. Scientists are still nutting out whether eyespots fool predators into thinking that the prey animal's head is there, or whether they are just conspicuous targets like the lizard and butterfly tails above. Nevertheless, the possibility of some sort of false-head illusion has been suggested for many of animals.

Hairstreak butterflies (*Calycopis* spp.) don't just have eyespots at the tips of their wings, they also have small wiry-looking tail tips that look like false antennae. Even jumping spiders, which have excellent eyesight and are known to be some of the smartest spider species, can be fooled into attacking these apparent 'false heads' instead of the butterflies' real heads. Some banded sea snakes (*Laticauda* spp.), also called sea kraits, have odd-looking tail tips. Instead of tapering off to a slender point, their tails have a bulbous and flattened tip, about the same size and shape as the snake's head. The flattened tail tips also have similar colour patterns to the snake's real head. This false head is apparently convincing enough to have fooled some sea snake researchers into thinking that a snake was looking outwards, when in fact the real head was hidden away, probing the rocky coral reef in search of food.

Once these brightly coloured tails or false heads have worked their magic and a prey animal has narrowly survived an attack from a predator, it's a good idea for that prey animal to use the second chance they have been given to make a hasty escape. On the move, there are even more tricks to be played.

Darting and dazzling

In the previous chapter we met venomous coral snakes and their colourful mimics. Their bold conspicuous colours are arranged in thin bands that run around the body of the snake. There are many other snake species that have conspicuous bold lateral stripes, but unlike the coral snake, they aren't toxic, nor do they appear to be mimicking any other toxic species. Scientists have noticed that banded snakes seem to be fast-moving species, whereas more slow-moving snakes have different colour patterns. In the same vein, scientists from the University of Uppsala found that in a population of *Vipera berus* snakes, males that had conspicuous zigzag stripes had higher survival rates and fewer predator attack scars than more plain-coloured black males. Both findings fit nicely with an idea called 'dazzle motion' that has been floating around for some time. Specifically, it is thought that bold stripes on fast-moving prey can confuse a predator and make it harder for them to accurately track or locate their moving prey.

You've probably experienced a similar illusion if, like me, you choose the movies you watch based on the likelihood of explosions, car chases and roundhouse kicks. There's always that one scene where the bad guy's car is rocketing down the highway, with one of his goons hanging out the back window, guns blazing. Amid all the action, you can't help but notice something strange about the shiny spokes on the alloy wheels of the getaway car. Instead of being a spinning blur they seem frozen in time. Every now and again it looks like they are slowly turning backwards. The same illusion happens in that scene where the Black Hawk helicopter takes off just before the entire army compound explodes into smithereens, yet you still notice that, at the right speed and the right angle, the blur of propellor blades seems to slow to a halt, while the helicopter stays floating in mid-air.

These illusions are a result of our eyes and minds interpreting the incomplete visual information presented to us by film footage. A run-of-the-mill piece of film footage is captured at around 24 frames per second. This is fast enough for our brains to interpret those 24 images – projected on screen one after the other – as smooth movement. This is also how animation works. As little as 12 frames per second can be sufficient for us to believe that hand-drawn images or sculpted lumps of clay have come to life on the silver screen.

This all works fine until we have fast objects, like spinning wheels and propeller blades, whose movement can't be tracked sufficiently with 24 frames a second. Suppose that the alloy spokes on that car wheel are spinning at precisely 24 revolutions per second. In one frame of footage, a wheel spoke may be captured pointing directly upwards, by the time the next frame of footage is captured the wheel has completed one rotation and that same wheel spoke is pointing directly upwards again. Continue at this pace, and the wheel is always captured on film in the same position, so you end up with a piece of footage where the wheel appears to be motionless. Slow the car down a fraction, so the wheel spokes do slightly less than one rotation every frame, and you will get a piece of footage where the wheel seems to be slowly turning backwards. Speed the car up and the wheel seems to crawl forwards.

Animal eyes don't take in infinite amounts of visual information. Just as a film camera can only capture so many images per second, our eyes can only process so many images per second. This is referred to as our 'critical flicker frequency' (CFF), which for humans is about 60 Hertz (Hz), meaning that our eyes process images somewhere around 60 times a second. Think about whenever you watch a fast-moving object, like the spinning blades of a ceiling fan. We don't perceive each individual blade of the fan moving, instead the blades blur together into a vague circle. The fan moves too fast for our eyes to track when we're only taking in 60 snapshots a second. Humans aren't particularly great when it comes to visual speed, and other animals can process images much faster. For a housefly, whose CFF is around 270 Hz, those spinning fan blades may be easy to track, like a gently turning windmill. The award for slowest vision in the animal kingdom currently goes to the giant African snail (*Lissachatina fulica*), which is only able to process visual information around once every two seconds.

Just like wheel spokes and fan blades get muddled up when their fast movement is processed using limited frame rates, the different coloured stripes on a banded animal, like a stripy snake, could also get muddled up depending on how fast it is moving and the CFF of the eyes watching the snake. Scientists who study banded snakes, and who make a living chasing live snakes around the bush, have described how the bands of colourful snakes start to blend, obscuring the snakes' movement or changing their apparent speed. Scientists from Newcastle University tested this in a praying mantis, *Sphodromantis lineola*. Praying mantises have excellent vision for detecting fast-moving prey. When scientists presented them with uniformly coloured moving targets on a computer screen, they would successfully strike. But when the artificial targets were fast-moving and stripy, the mantises couldn't quite figure out what to do and often wouldn't respond to the moving prey at all. In another study, scientists determined the CFF of bird eyes and then calculated that when certain banded snakes were moving at top speed, their colour patterns probably couldn't be distinguished by birds.

For dazzle motion to work in this way relies on the speed of the prey, plus the size of their stripes, syncing with the CFF of the predator's eyes. There is another way that motion dazzle might work that doesn't require this perfect confluence of conditions. And it has to do with the fact that busy contrasting patterns are just confusing in general. Look at the image on the following page or, better yet, keep reading a paragraph opposite it while noticing what the pattern seems to be doing. Every now and again it should look as if the circle is gently rotating. When you look directly at the image the movement seems to stop. This is called the 'peripheral drift illusion'. When our eyes and minds try to make sense of these high contrast patterns using our peripheral vision, they perceive movement that isn't there.

There is a style of abstract art called Op Art (as in, optical art) that uses busy, often black and white abstract

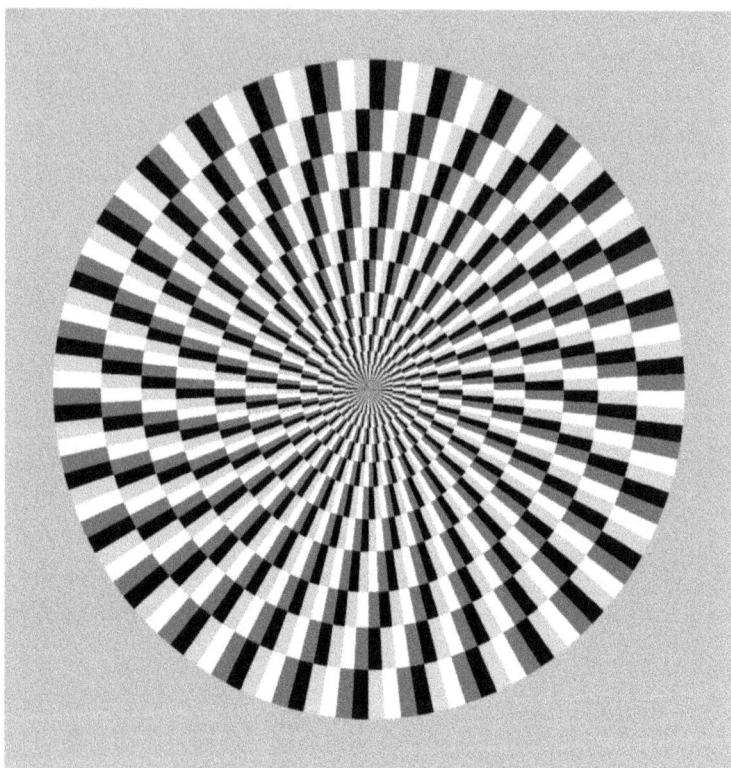

The peripheral motion illusion

patterns that create the illusion of movement in static images. Viewing a piece of Op Art can be a jarring experience as you realise you are unable to focus on a single point in an image, that your focus is constantly being pulled in all directions by jagged lines and shapes. Now imagine a high contrast and stripy animal darting away into your periphery. Could their striking patterns start to become a confusing mess? And could this jumble of information confuse a predator just enough to give the prey a slight advantage?

The possibility of dazzle motion as an anti-predator strategy has been suggested for all kinds of stripy animals, be they snakes, fish, lizards or zebras. A similar phenomenon has been described in certain birds, and in butterflies and other insects, where their brightly coloured wings seem to flash different colours as they flutter away. This 'dynamic flashing' display of multicoloured wings apparently makes it hard for predators to accurately calculate the trajectory of their moving prey. We still don't know exactly what is going on in the eyes and minds of the bamboozled predators. We don't even fully understand why optical illusions like the peripheral drift illusion work. But they quite obviously do. Most evidence for dynamic flashing and dazzle motion comes from studies where humans are asked to play simple video games where they must try to click on little striped shapes zipping across computer screens. These kinds of experiments are understandably much easier to conduct than ones that try to get inside the mind of a predator chasing live snakes through the undergrowth.

Even though we don't fully understand how dazzle motion works, we seem rather confident that it does.

So confident that during World War I over 2300 British ships and over 1200 American ships were painted in elaborate light and dark dazzle markings to thwart ongoing attacks by German submarines. Since submarine gunners aimed their torpedos ahead of moving enemy ships, dazzle markings should, in principle, have inhibited the ability of the gunners to accurately gauge the speed of the moving target. More recently, entrepreneurs have designed striped wetsuits and surfboards in attempts to create shark-proof surf gear. I don't know how these things have fared in field tests, nor do I think they will come with a customer satisfaction guarantee.

Recent research has refined this potential method of shark deterrent and tested whether brightly lit patterns might deter shark attacks. Since sharks would be viewing a surfer from below, the surfer or surfboard would be a dark silhouette on the surface and any patterns on the surfboard or wetsuit would probably be hard to spot. Australian scientist Laura Ryan and her colleagues tested whether bright lights on an object would work to deter sharks. They tested this in the most real-world setting possible. They constructed a seal-shaped dummy with bands of bright blue LED lights on the underside and towed in behind a boat in great white shark territory off the coast of South Africa. The unlit seal models were frequently followed and attacked by great white sharks, but the model seals with brightly lit stripes survived unscathed.

If all else fails, play dead

Imagine you are a prey animal being chased by a predator, and all your hiding, deflection and escape strategies have failed. You're face to face with a snarling beast and within moments of death. You may as well roll over and die, right? Nah, it wouldn't be that simple, would it? Though, when face to face with an attacker, some animals do seem to roll over and die but (to the surprise of no one that has read this far into the book) it's not what it seems.

North American opossums (*Didelphis virginiana*) are famous for a bizarre behaviour whereby, when they are threatened by a nearby predator, they suddenly play dead. They freeze on the spot and roll onto their sides, with their jaws hanging open in a pained expression. Though opossums are renowned for, you know, 'playing possum', all kinds of animals are known to play dead in response to predator attacks, including insects, spiders, birds, rabbits, snakes, frogs and lizards. This bizarre behaviour called 'death feigning' is a bit of a head-scratcher.* You would think that a predator faced with a suddenly immobile prey would be chuffed about the obliging behaviour of their next meal. However, death feigning seems to have the opposite effect. In one study, domestic cats overlooked immobilised quail in favour of mobile ones. Another study showed that ducks could survive attacks from foxes by death feigning.

---

* Also called 'thanatosis', after Thanatos the Greek deity of death who, funnily enough, also served as inspiration for Thanos, the famous Marvel villain and the focal topic of another long, meandering and unnecessary footnote. Anyway, how's your day been?

One potential explanation for why this works is that the predators are genuinely fooled into thinking that their prey is dead and are deterred from eating it in case it is diseased or infected. Some prey animals give a bit of extra flair to this necrotic performance. When playing dead, opossums release a green putrid-smelling fluid from their anuses. Other animals drool, defecate and urinate as if to make themselves as distasteful as possible. Hognose snakes (*Heterodon* spp.) pull out all the tricks. They roll onto their backs exposing their white underbellies, coil into spirals and writhe as if in agonising death throes. Their mouths open wide, their tongues hang limply and they ooze foul-smelling liquid from their cloaca. I can imagine that such disgusting displays would be enough to deter many predators regardless of whether they believe their prey is dead or not.

If rolling over and playing dead seems a bit anti-climactic as a last-ditch attempt at using deception for survival, fear not; some of the most spectacular displays in nature come from animals that take a much more bombastic approach to bluffing their way out of other animals' jaws.

One last almighty bluff

Peacock butterflies (*Aglais io*) are what a birdwatcher might refer to as 'little brown jobs'. This colloquialism describes the many different plain-Jane bird species that are small, brown, inconspicuous and often near-impossible to tell apart. As a scientist who studied insects (but was admittedly never very good at telling one species of insect from another)

I found 'little brown jobs' a useful term to apply to all the other small, brown and inconspicuous-looking critters I came across, be they cockroaches, crickets, butterflies or beetles. I've never had the opportunity to spend any time studying butterflies in detail so, to this untrained eye, peacock butterflies fit this description. Which raises the question of why they are called peacock butterflies. No, it's not an ironic dig at their dull appearance, there is another side to the peacock butterfly that only gets revealed when the going gets tough.

At rest, peacock butterflies sit with their wings held together pointing upwards. The visible undersides are a pale mottled brown that matches the plain colours of their furry heads and bodies. But when a predator approaches and is just about to attack, peacock butterflies flick open their wings, revealing a gaudy flourish of colours. The topsides of their wings are bright red with dark black bands lining the front edges, and white and blue eyespots on each of the four wing tips. Along with this sudden flash of colours, they emit hissing noises and ultrasonic clicks.

Scientists have given these kinds of behaviours a fancy name: 'deimatic displays'.* Though a more common and intuitive term is 'startle displays'. The idea is that these spectacular displays are so sudden and unexpected that they startle, intimidate or frighten off the attacking predator. At the least they should cause a predator to hesitate, giving the prey a sliver of time to make a quick exit. Numerous studies have shown that peacock butterflies' elaborate

---

* From the Greek word for 'to frighten'. Just as in Deimos, the Greek God of Terror. Those Greeks were a morbid bunch, eh?

displays are alarming enough to deter small birds and rodents from attacking what is otherwise a harmless and undefended butterfly.

Scores of generally harmless creatures use bold and brilliant displays to try to bluff their way out of becoming caterpillars' lunch. Walnut sphinx moth (*Amorpha juglandis*) often emit a high-pitched whistle. To humans, it sounds a bit like a small squeaky toy, but it's enough to deter small predators like red-winged blackbirds. Some fish and squid have been seen giving off bright bioluminescent flashes when being attacked by elephant seals. Australian frilled-neck lizards (*Chlamydosaurus kingii*) famously have an elaborate fleshy collar that they fan out to display a red-orange shield circling the bright yellow skin of their open mouths. When fully expanded, the frill can be up to six times as wide as the lizard's head. Blue-tongue skinks (*Tiliqua* spp.) open their mouths and extend their large and bizarrely coloured tongues. Mountain katydids (*Acripeza reticulata*) look like small dark grey balls with gangly legs. When under attack, they lift their wings, revealing abdomens covered in bright red bands and vivid blue spots, plus some oozing droplets of bitter chemicals for good measure. Thanks to their colour-changing wizardry, cephalopods don't even need to move to show off sudden startle displays. The European cuttlefish's display involves quickly turning bright white with a dark ring around the rim of their bodies, and two black eyespots appearing on their backs. They can even target their display so that the pale white deimatic colours show on the side of the body pointed towards the predator, while the opposite side stays camouflaged.

Praying mantises are famous for their startle displays. When under threat, they rise on their hindlegs, stretch their forelimbs wide and fan out their wings. In some species, they wave their bodies side to side and reveal bright colour patches on their hindwings and the inner surfaces of their forelimbs. Some add rasping noises to the display by rubbing their wings against their abdomens. The internet is littered with videos of praying mantises frantically displaying to attacking birds, curious cats and irksome people brandishing smartphone cameras. Insect photographers worldwide (myself included) have prodded many a mantis trying to capture the glory and horror of their technicolour defensive displays. And as a result, I have been on the deserving end of some very sharp and painful praying mantis strikes.

Because these are last-ditch efforts to deter predators, deimatic displays are usually paired with some other sort of strategy, like camouflage. Dead leaf praying mantises (*Deroplatys* spp.) are a beautiful example of this. At rest, these stoic creatures look astoundingly like a crumpled brown leaf, but when threatened, they stand tall and reveal a burst of vivid burnt orange on their undersides. Their raised hindwings are lined with bold black and white stripes. On their forelimbs two black spots look like piercing black eyes, and glistening white spines look like piercing teeth around a gaping maw.

There's no evidence to suggest that this elaborate display is supposed to mimic a face with giant gnashing jaws. Most deimatic displays don't look like anything specific at all. The sudden display of garish patterns seems enough to do the trick. Though eyespots do seem to add a little extra freak-factor to these scare tactics. When scientists experimentally removed the eyespots of peacock butterfly wings, their displays were less intimidating to approaching predators. If you put yourself in the mindset of being a predator stalking a seemingly harmless butterfly who suddenly comes face to face with an enormous pair of eyes staring straight back at you, it's easy to imagine how these startling eyespots could work. Though there still isn't any rock-solid evidence that such eyespots resemble real eyes. Even if they don't, they at least add enough extra pizzazz to the display to make a predator think twice.

One animal has a startle display that is so elaborate, it is hard to imagine how it *couldn't* have evolved to fool a predator into thinking another predator is staring right back at them, and it has one of the most famous deimatic displays ever described. The caterpillar of the South American hawk moth *Hemeroplanes triptolemus* is large for a caterpillar, about the length of your thumb, and is completely harmless. Though their elaborate deimatic display makes you seriously doubt this. They bend over backwards, pointing their heads towards the ground and start inflating the first few segments of their bodies. As they expand outwards, small slits along their sides open and two dark black patches emerge like giant blinking eyes. Their inflated segments take on the uncanny form of a snake's head complete with white spots on the snake's false eyes, giving the almost unbelievable impression of light sparkling off the surface of glossy black snake eyes. Even knowing that this is a harmless caterpillar, it's hard not to sense the threat of a viper poised and ready to strike.

If and when to defend

Keep in mind that not all intimidating displays performed by animals are empty threats. Think about our old friend, the tiger. Imagine a fully-grown tiger, with glaring orange and black stripes, arching its back, baring its teeth and snarling in your face. This tiger isn't bluffing. It's sending a bluntly honest message about its ability to rearrange your organs. The same could be said for the colourful banner of a venomous cobra's hood. Or the chilling snare of a rattlesnake's tail. On the other hand, a small butterfly flapping its wings in your face doesn't have the bite to match its bark. The predator must be fooled by the butterfly's bluff for this strategy to pay off.

Animals that use these deceptive displays need to make tricky decisions about if and when to use them. Imagine a scenario where a small and perfectly edible praying mantis is camouflaged against some leaves on the ground. It becomes aware of a large bird moving closer and closer. The closer the bird gets, the more vulnerable the mantis is to attack. So, it could make sense for the mantis to make the first move and put on the most flamboyant startle display it can muster. But the camouflaged mantis can't know for certain whether it has been spotted by the bird or not. If the bird hasn't seen the camouflaged mantis, then a pre-emptive startle display might simply act as a bright waving flag, directing an otherwise oblivious predator towards their next meal. Alternatively, a startle display could be used at the very last minute, when the mantis is absolutely certain that they are being attacked. This is an inherently risky strategy as it could mean holding off until the moment the predator strikes or even after an initial strike.

Some colleagues and I studied this conundrum in Australian praying mantises. We used a very sophisticated and super-sciencey approach of collecting some praying mantises and poking them until they flipped out and showed off their startle display. Sometimes we would wave our hands in front of them without touching them, other times we poked them with a little stick, and other times we would give them a gentle pinch with a pair of forceps. This was intended to simulate different types of threatening encounters, ranging from having something big and scary-looking nearby, through to direct physical contact. Across three different species of praying mantis the pattern was the same each time – the mantises rarely reacted to a waving hand, but were much more likely to respond when poked, and even more likely to show a startle display when pinched with forceps. It's as if the praying mantises waited until the very last minute – when they were under a simulated predator attack – to play their startle card.

While I'm making light of the seemingly simple methods we used, it's an approach that other researchers have used to study deimatism. Scientists have poked, prodded and pinched all sorts of animals to see what it takes to provoke a startle response, it's just that when they write their formal scientific papers they don't say 'we poked them till they flipped out', they say something clever like 'we examined their responses to looming artificial stimuli'. Some older studies took more direct approaches by simply putting insects in cages with birds and monkeys and watching what happened. In most cases, a similar pattern emerges. Spotted lanternflies, mountain katydids and several praying mantis species all seem to wait until a predator has started attacking, then use startle displays as a last-ditch effort to bluff their way out of a dicey situation. If they are already getting attacked, they may as well kick up a fuss.

While startle displays seem like a risky survival strategy, there is another animal that uses a bold bluff to dice with death in a completely different way. Like the animals at the beginning of this chapter, it uses the magic of misdirection to deflect the attacks of predators. But there's a twist (literally and figuratively): it uses misdirection to flip the script and turn their predator into prey. If you need a break from reading to go and make a cup of tea, now's the time to do it. In the next chapter, nature's tricksters get brutal.

5

Death by deception

There are three types of people: those who look up while walking through the forest, those who look down and the rare and mysterious few who look straight ahead. I, like any sensible person, look down. Hidden among the rocks and low shrubs are all manner of delightful creatures to be spotted; millipedes, grasshoppers, skinks, slugs, the list goes on. The added bonus of looking down is that it helps you spot any protruding rocks or logs in your path and helps you avoid stepping on the odd basking snake or fresh animal droppings. Up-lookers, while I'm sure they are lovely people, are a bit silly. They wistfully gaze into the canopy and mumble to themselves something about the lovely hues of autumn leaves, until they trip on a fallen branch and land flat on their face. An up-looking twitcher,* busy peering through their narrow binoculars, with a leather strap poised to garotte the tumbling leaf-nerd, is a deadly combination. I'd suggest avoiding these people as bushwalking companions, or at least be ready to ring for an ambulance at a moment's notice. The straight-ahead lookers are unusual creatures that doggedly march

---

* Passionate birdwatcher

forward, unswayed by the colours and complexity of the canopy above and the understorey below. They scan the horizon looking for goodness knows what, glancing at the bare surfaces of tree trunks. Who knows what goes on in the minds of these mid-range gazers? Thankfully they are rare and appear generally harmless. In the early days of my scientific career, I was, for professional reasons, forced to adopt the perspective of a forward-looking bushwalker when I studied a particular type of insect that lived on the bare surfaces of tree trunks.

If you're in the business of spotting animals, then looking either up or down is a good strategy. In forests, most animals make their homes in either the understorey below or the canopy above. If you look straight ahead, you will find yourself staring at tree trunks which, it might seem, are strange habitats for animals to live in. They are exposed surfaces that don't offer much in terms of food, and they lack the protective cover offered by the leafy canopy above or the rocks and leaf litter on the ground below. Comparatively few animals tend to make tree trunks their permanent homes. Instead, tree trunks are like forest highways that connect the canopy with the understorey. Trails of ants weave up the length of a tree trunk as colonies climb up searching for food before descending to bring the food into their subterranean nests. Flying insects, like flies, beetles and moths, will rest on tree trunks temporarily but won't hang around for very long. The few animals that do eke out a full-time existence on tree trunks often use these highways as an ever-replenishing food supply.

Such were the insects that I was searching for in the tropical forests of Australia. Tree-running mantises

(*Ciulfina* spp.) are small slender praying mantises that use many of the camouflage strategies we've covered in the book so far. They are mottled brown, matching the bark beneath them, with dark stripes that run across their heads masking their tell-tale bulging eyes. When lying flat against smooth bark, they are incredibly hard to spot and the best way to find them is to run your hand across the bark hoping to scare one into moving. If one does, it will dart across the bark surprisingly fast. The flash of movement breaks the illusion of camouflage and reveals their position. Like many of the other animals that live fulltime on tree trunks, they are opportunistic predators that cruise around searching for small insects traversing the tree trunk highway network. There are long-legged spiders the colour of bark and lichen that wrap themselves around the curved bark surface waiting in ambush for a passing meal. There are geckoes that shelter under bark sheets during the day and come out to hunt at night. And in south-east Australia there is a predatory bug that makes its home on the thick trunks of eucalyptus trees, and hunts its prey in a way that is unlike any other animal on the planet.

Up until this point, we have mostly been exploring how animals and plants can use deception to protect themselves. This chapter, however, focuses on how creatures use deception to hunt their prey. As we'll see, there are plenty of creatures that use clever lures and elaborate disguises that entice their prey to walk, fly and swim right towards their hungry mouths. The feather-legged assassin bug (*Ptilocnemus lemur*) lives up to its name in both appearance and nature. The lower section of each hind leg is covered in a dense coat of thick, wiry hairs. Picture an enormous pair of knee-high furry legwarmers. Now, picture those legwarmers on a small yellow and black bug about the size of your fingernail. It doesn't paint the most intimidating picture of a formidable predator, but this assassin bug uses its spectacularly adorned legs to take down some of the most dangerous animals in the insect world – ants. As we've already learnt, ants are formidable fighters, and many predators avoid them altogether, but feather-legged assassin bugs have evolved a complex arsenal especially tailored to taking them down.

A juvenile assassin bug will find a trail of ants meandering up a tree and stand a few centimetres away from it. While many other small animals wouldn't dare get this close for fear of becoming ant-food, the assassin bug does something surprising: it starts waving at the ants. It raises a long feathery hindleg and starts wiggling it up and down. Ideally, this attracts the attention of an ant that will depart from its path along the tree trunk to investigate what this strange wiggling thing is. As the ant approaches the assassin bug continues waving its leg until, eventually, the ant strikes – biting down on the bug's leg.

In any other circumstance, this would be game over for the bug, but the feather-legged assassin bug has a bizarre trick up its sleeve. Once the ant has grabbed hold, the bug pivots around the leg joint and jumps right on top of the ant's head. From this vantage point, the bug extends its proboscis (long needle-like mouthparts) and stabs it into the base of the ant's head, right where the ant's rigid exoskeleton softens to form a moveable neck joint. With the bug's leg still held tight by the ant's powerful jaws, the bug injects venom that paralyses the ant. On subduing the ant, the bug starts injecting digestive fluids that turn the ant's body cavity into a delicious soup. Somehow, the thick feathery coating of the hindlegs protects the bug from the crushing bite of the ant. Using an incredibly bold bait-and-switch ploy, feather-legged assassin bugs elicit a predatory strike from an ant, allowing them to perform a sort of miniature jiujitsu move that turns the hunter into the hunted. The technique is so effective that feather-legged assassin bugs can take down ants that dwarf them in size.

There are many different species of feather-legged assassin bug, and this strange predatory tactic has only been studied in juveniles of the Australian *Ptilocnemus lemur*. But this might be only one of many tricks that these bugs use to take down ants. There are strange records of adult feather-legged assassin bugs using a completely different, and very mysterious, predatory tactic. In a report from 1911, Dutch naturalist Edward Richard Jacobson tells the story of a similar assassin bug, *Ptilocerus ochraceus*, that he found in Java, Indonesia. He watched them on posts and bamboo poles, where they would stand beside ant trails. Unlike the bugs in the story above, these assassin bugs didn't wave their legs to get the attention of ants; instead, the ants seemed to walk towards them of their own volition. Once an ant got close, the bugs would rear up to expose a 'curious tuft of yellow hair' on their undersides.* The ants, apparently mesmerised, walked right underneath the bug,

---

* Edward Richard Jacobson (1911) Biological notes on the Hemipteron *Ptilocerus ochraceus. Tijdschrift Voor Entomologie* 54, p. 177.

and placed their mouthparts against this tuft of hair as if they were tasting it. The assassin bug then carefully placed their forelegs on the ant's head, readied their proboscis, and stabbed the ant in the same vulnerable patch of soft exoskeleton behind the ant's head. As far as Jacobson could tell, the bugs were secreting some sort of chemical from their undersides, that lured and sedated the ants.

Jacobson's story starts to wander even further into the incredible, as he describes swarms of these bugs in the thousands, aimlessly drugging ants and covering the ground at his feet an inch thick with ant carcasses. He does admit that this is a phenomenon he never witnessed again. And, to my knowledge, no one has ever seen this or any other feather-legged assassin bug leaving behind swathes of paralysed ants like this. Nevertheless, it does seem that adult feather-legged assassin bugs somehow draw ants towards them and appear to chemically sedate them before killing them.

The tuft of yellow hairs that Jacobson refers to is a strange structure called a trichome. All feather-legged assassin bugs seem to have one. It sits on the underside of the bug, just beneath where the legs join the body. Under a scanning microscope it appears as a beak-like gaping maw, fringed with dense hairs and smattered with pores. Behind this lie several small internal glands that, it is assumed, must be the source of the ant-sedating chemicals. Beyond this, we don't really know much about the trichome and what it does. Matthew Bulbert, the scientist who described the bait-and-switch manoeuvres of juvenile *Ptilocnemus lemur*, has also observed ants being drawn to and somehow mesmerised by the trichome of adult Australian feather-legged assassin bugs. We don't know what these chemicals are or how these bugs have evolved such intricate strategies to take down their prey. It's fair to assume that these small, fantastically shod assassin bugs harbour many more mysteries waiting to be uncovered.

Spider tails, worm tongues and fly traps

Earlier in the book, we explored how camouflage can help ambush predators hide from any unsuspecting victims that might wander within the predators' reach. Some ambush predators, however, don't seem content to wait patiently and have evolved ways of luring their prey closer without breaking camouflage. Many snakes, particularly vipers, use a technique called 'caudal luring', where the very tip of the tail is wriggled about like a worm to attract the attention of small animals. The Amazonian two-striped forest pit viper (*Bothrops bilineatus smaragdinus*) hunts in treetops where its green colour seems to help it blend in with the green leaves of the forest canopy. The very tip of its tail, however, is a brown or creamy-white colour. When hunting, the vipers coil on a branch and wriggle their tail tips alluringly across their bodies and underneath their poised and waiting heads. The Iranian spider-tailed viper (*Pseudocerastes urarachnoides*) has a more elaborate lure at the tip of its tail: a bulbous tip fringed with long flexible spines which, when waved about, creates the illusion of a spider crawling about on the ground. Scientists have observed birds swooping in to attack this supposed spider only to be caught by the rapid strike of the patient snake.

Some frogs use a similar strategy, apart from the fact that they don't have tails and instead waggle one of the toes on their hindlegs to attract the attention of a nearby animal.

Some of the most famous examples of lures in the animal kingdom are those used by anglerfish. Anglerfish (sometimes called frogfish) are masters of camouflage. Some species look like calcareous algae or sponges, sitting motionless on the seafloor. Others are intricately festooned with branching fronds, helping them blend in among seaweed. In some species, the first spine of the dorsal fin is modified to form a forward-facing spine with a lure at the end. This lure might look like a vague blob, a frilly tassel or a curling worm-like shape. The idea is that the lure is dangled in front of the anglerfish's face and when another curious fish comes to investigate, the anglerfish strikes, rapidly opening their prodigious jaws and sucking their unsuspecting prey into their mouth. Deep-sea anglerfish can have glowing bioluminescent lures. The deep-sea anglerfish (*Oneirodes* sp.), introduced in the first chapter as the blackest animal in existence, is thought to have ultra-black skin so that they can activate the bioluminescent glow without illuminating and revealing their own faces, waiting eagerly behind the dangling lure.

Other animals use a more direct approach when enticing animals towards their hungry mouths by using their tongue as a moving lure. Alligator snapping turtles, (*Macroclemys temminckii*), have a strange appendage on the tip of their tongue that unfurls and wriggles about like a worm or grub. These turtles sit motionless on the riverbed and open their mouths wide. The bright red worm-like lure wriggles back and forth, attracting the attention of nearby fish. Any fish that is duped into attacking the lure, thinking that it is a juicy grub, is easy prey for the turtle that snaps its jaws down on the unsuspecting fish. Mangrove saltmarsh snakes (*Nerodia clarkii compressicauda*) also lure fish with their tongue, but do it by dangling their tongue at the water's surface where it draws the attention of small fish looking for food.

Aquatic garter snakes (*Thamnophis atratus*) elaborate on this visual stimulus by using their tongues to create a vibrational lure. These snakes slither among rocks near rivers and streams. While perched on a rock, they will lower their heads until almost touching the water. Then they will poke out their tongue and rapidly quiver it on the water's surface. The ripples this creates are thought to resemble the vibrations made by small animals, like insects, struggling on the surface. Any fish, frogs or salamanders that come looking for an easy meal quickly have the tables turned on them by the hungry snake waiting above.

Birds can use the same strategy. The snowy egret (*Egretta thula*) wades through shallow floodplains hunting small fish and has been seen lowering its beak into the water and creating enticing ripples by flicking its tongue. Just like ripples can be sent through the water, other animals send ripples through spider webs to attract the attention of the web's resident spider.

Another Australian assassin bug, *Stenolemus bituberus*, specialises in hunting spiders. It does so by gently plucking the silk threads of spider webs. When the spider comes wandering over to see what is making these strange web vibrations, the assassin bug attacks from above, stabbing its sharp proboscis into the back of the spider's head.

Luring prey is a genius hunting strategy that takes away the effort of finding and chasing down unsuspecting prey. With just the wriggle of a toe, a waiting predator can bring their prey right to them. Lures like wriggling tongues, toes and tails can be switched on and off. Other animals have more passive means of luring prey that are permanently switched on. Many orb web spiders have bright yellow and black stripes that run across their backs. While they sit motionless in the centre of their spiderwebs, these bright contrasting colours attract the attention of flying insects that use similar bold colour patterns to find flowers. Should a curious pollinating insect get too close to these brightly coloured spiders, they may unwittingly fly straight into the spider's sticky web.

Jewelled spiders (*Gasteracantha fornicata*) also have bold yellow and black bands that run across their backs. Australian scientist Thomas White noticed that these spiders tended to sit in the centre of their webs at a slight angle, so that their bold parallel stripes ran diagonally. He found that spiders sitting at an angle caught more prey than spiders whose stripes were oriented horizontally or vertically. White suggested that this tapped into some aspect of insect eyes and minds that made the spiders less recognisable as a potential predator. In the first chapter, we encountered Australian crab spiders that, unlike other crab spiders, don't camouflage against white or yellow flower petals. Instead, they reflect bright ultraviolet light. The contrast between the UV-reflecting spider and the UV-absorbing flower petals piques the interest of pollinators, who come to visit the flower only to be attacked by the waiting spider. Rather counterintuitively, scientists have shown that flowers with conspicuous UV-reflecting crab spiders on board attract more pollinators than flowers with camouflaged spiders.

Animals that we don't generally think of as predators use similar tricks to lure prey. Glow-worms use bioluminescence in dark caves to lure prey into a silk trap. These glow-worms are not true worms, but the larvae of flies (genus *Arachnocampa*) that hang from the ceilings of caves in Australia and New Zealand. From their high perches, they dangle long threads of silk dotted with sticky mucus. At night, the worms' glowing blue bodies attract flying insects that quickly become entangled in the sticky mucus trap. Siphonophores, strange marine creatures that are essentially like floating colonies of jellyfish, have stinging cells with glowing red spots. It's possible that this tantalising light display tempts nearby animals to swim into the network of stinging tentacles.

Carnivorous plants appear to play the same game. The hinged leaves of snapping Venus flytraps reflect bright UV light in the centre, which contrasts with the UV-absorbing rim and elicits the attention of flying insects. Pitcher plants have modified leaves that curl to form a deep well. The plant secretes digestive fluids that pool at the bottom of this well to trap and kill any small insects that fall inside. Scientists have shown that insects are attracted to the deep red colours that line the entrances of the pitcher plant trap. Pitcher plants from North America and Borneo release chemicals similar to the scents of flowers and fruit. Animals following these scents, imagining that they might encounter a nutritious meal at the end of these chemical trails, eventually stumble into the pool of digestive fluids inside the pitcher plant's trap.

Prey-luring uses a simple strategy: create a signal that promises a false reward to an animal. This is usually something associated with food. The bright colours of spiders pique the interest of pollinating insects that associate bright contrasting colours with nectar-filled flowers. The wriggling lures of assassin bugs, snakes and snapping turtles trigger interest from animals that feed on small wriggly things. To us, they might not look at all like flowers or wriggling worms, but it doesn't matter. All that matters is that they are enticing enough to make a prey animal move towards a predator instead of away. It's possible that in some cases, like that of predatory glow-worms, a passing resemblance to food might not even be necessary. Anyone that has ever sat outside on a warm evening and listened to the crackling noise of insects being zapped out of existence by an electrified light trap knows many flying insects are hardwired to be drawn toward bright lights. In scientific parlance, these prey-luring strategies are described as 'sensory exploitation' as they work by tapping into biases in animals' sensory systems. Other tricksters take things a little further and rather than presenting 'good enough' lures to entice their prey have evolved much more elaborate and persuasive resemblances.

Aggressive mimicry:
honey traps and singing sirens

In chapter 2, we encountered spiders and caterpillars that seem to masquerade as glistening bird poo to protect themselves from predators. For spiders, this doesn't just help protect them from predators, it also helps them lure prey. Glistening white bird droppings might not seem very appealing to us, but they are gourmet treats for flies. *Phrynarachne ceylonica* crab spiders in the tropical forests of China are considered 'aggressive mimics' of bird droppings and attract flies, just like real bird droppings. Unlike the lures described earlier, aggressive mimics don't just have a vague or fleeting resemblance to something – they can appear uncannily convincing. Bird-dropping spiders like *P. ceylonica* come complete with contrasting patches of white and brown, shiny wet-looking surfaces and rough poopy textures.

Aggressive mimicry works on the same principle as sensory exploitation; the aggressive mimic should resemble something rewarding to a target prey animal. Again, and in the case of bird-dropping spiders, resembling something nutritious is a surefire way to draw in hungry animals, but there's more than one way to skin a cat, or catch a moth, whatever tickles your fancy. For example, bolas spiders lure moths, but the moths aren't tricked by the promise of food, they are tempted by the scent of romance. Bolas spiders (*Mastophora* spp.) get their name from the ingenious webs that they build. Instead of constructing an elaborate net to entangle prey, their webs are reduced to a single thread with a heavy glue droplet dangling at the end, just like the traditional bolas hunting weapon that consists of lengths of rope with weighted ends. With one leg, the spider twirls the silk thread around in large circles and traps any moths that fly into the path of the spinning bolas. Rather than just sit around idly twirling silk in the hope that it catches on to something, bolas spiders also release chemicals that smell just like the pheromones of female moths. Male moths flying around at night on the hunt for females are fooled by this chemical mimicry and fly straight into the spider's spinning trap.

An Australian predatory katydid lures its prey by singing a romantic tune irresistible to cicadas. At night, male cicadas call for females using loud, sometimes ear-splitting, chirps. Nearby females, if they are receptive to the boisterous blaring males, will respond with a more modest clicking sound. The males then go searching in the darkness for the source of the clicking sound. The calls continue back and forth in a coordinated duet; males call and the females respond. All the while, the male continues triangulating in on the female clicks to find his mate. The predatory katydid *Chlorobalius leucoviridis* hacks into this call-and-response song by mimicking the clicks of female cicadas. Male cicadas thus find themselves calling back and forth with deceptive katydids, inching their way towards deadly singing sirens.

Certain fireflies use a similar trick to lure in males as a quick and easy meal. The flickering constellations of glowing fireflies are the way that males attract females. Each male produces bioluminescence through a chemical reaction inside a specialised abdominal organ. The flashing patterns of their bioluminescence are unique to each species, so even among a swarm of different firefly types, females can seek out the distinctive flashing of their own kind. Females have their own light-producing organs that they use to signal their interest back to males. By directing their flashes towards a male of their choosing, the pair can home in on each other through the darkness of night and the swarms of competing fireflies. Female *Photuris* fireflies, however, exploit this communication channel to lure males of other firefly species as food. The females will respond to male mating flashes with their own seemingly receptive flashes, and they can modify their response flashes to manipulate the behaviour of several different prey species. The hungry female and naïve male flash back and forth, moving closer and closer, until the female pounces on the male, holds on tight with her legs and starts chowing down with her sharp mandibles.

A small coral reef fish called the bluestriped fang-blenny (*Plagiotremus rhinorhynchos*) cheats its way to snacks by using a rather unique and elaborate trick; they don't lure their prey by promising food or fornication, they trick them by offering a dodgy cleaning service. In the tropical seas of the Pacific and Indian oceans are small fish called cleaner wrasse (*Labroides dimidiatus*), popular inhabitants of coral reefs. They have bright black and blue stripes that advertise their underwater salons, like the bright stripes of a barber's pole. Cleaner wrasse swim around other much larger fish, picking off and eating small parasites. The oceanic rule of big fish eat little fish is set aside and a truce is declared as the larger fish even let cleaner wrasse swim right into their gaping mouths to clean their teeth and tongues. Both parties benefit, as the cleaner wrasse gets a meal and their clients are cleaned of parasites. This would all be wonderful and the fish could all swim home happily if it weren't for those dastardly fangblennies. These aggressive mimics have black and blue stripes, identical to cleaner wrasse. Unlike cleaner wrasse, they don't observe the underwater armistice – they swim right up to larger fish and, posing as cleaner wrasse, take a chunk right out of their sides.

Aggressive mimicry works a lot like Batesian mimicry, in that there are three parties involved: the model, the mimic and the unlucky dupe. In the example above, the cleaner wrasse is the model, the fangblenny is the mimic and the parasite-covered fish is the unlucky dupe that mistakes a fangblenny for a cleaner wrasse. The main difference between these two types of mimicry is that the outcomes are opposite – Batesian mimics trick predators into avoiding them, aggressive mimics trick prey into being attracted to them. Where aggressive mimicry differs from sensory exploitation is that aggressive mimics resemble a very specific and recognisable reward, whereas sensory exploiters offer a more generalised or ambiguous reward. At least, that's the idea. The truth is that telling the difference between aggressive mimicry and sensory exploitation is harder than it looks. Once again it comes down to the fact that assessing similarities from a human perspective is a poor approach to understanding what's going on in the eyes and minds of other animals.

The orchid mantis: a legendary flower mimic ... maybe

When I started my career as a scientist hunting for tree-running mantises hidden in the rainforests of northern Australia, what intrigued me about these insects was that they didn't fit the mould of what a praying mantis was supposed to be. You have probably found praying mantises in your garden, or at least have seen pictures of praying mantises, that are big and green with enormous round eyes either side of their triangular heads. We think of them as large and slow-moving insects; stoic creatures that sit camouflaged among the leaves, holding their spiked forelimbs at the ready, waiting to strike at any unlucky insect that wanders past. *Ciulfina* tree-running mantises are not like this at all. They are small, brown and fast. They use the cylindrical form of the tree trunk to hide from predators. When they spot something coming, they will dart around to the opposite side of the tree trunk at blink-and-you'll-miss-it speeds. The smoother the tree bark, the better, as they can run even faster.

At the time I started studying these praying mantises, we knew very little about their biology and behaviour. The more I learned about them, the more it became clear that we don't really know much about praying mantises at all. Overall, praying mantises are mysterious creatures. While famous for their good looks and their penchant for sexual cannibalism, we know close to nothing about the biology of most species and there are many, like tree-running mantises, that don't fit the stereotype of a stoic green garden visitor. As I was diving into the world of unusual praying mantises, there was one species that stood out above all others. It is the one that I introduced at the very beginning of this book; the impossibly beautiful orchid mantis, *Hymenopus coronatus*.

One of the earliest descriptions of orchid mantises in western literature comes from an Australian journalist, James Hingston. In 1879 Hingston was travelling through the then Dutch-ruled colony of Buitenzorg, or as it is now known, Bogor, just south of Jakarta. While staying at a homestead there, he was taken by his host to view a spectacle that he described in his memoirs as 'a flower, a red orchid, that catches and feeds upon live flies. It seized upon a butterfly while I was present, and enclosed it in its pretty but deadly leaves'.* I can only imagine what was going through Hingston's mind as he tried to comprehend what he was seeing. As far as he could tell, a flower had just sprouted legs and walked around right in front of his eyes. It was, I imagine, one of those moments that shakes the foundations of your reality, when you learn that something impossible may not be so. Since then, many other naturalists have tried to capture the surreal beauty of orchid mantises in tones almost as mythical-sounding as James Hingston's.

Books, blogs and biology lecture slides are awash with pictures of orchid mantises. Their stunning white and pink colours and charming resemblance to a flower blossom has earned them infamy as classic examples of aggressive mimicry. The flower-like guise of orchid mantises was believed to be so convincing that bees and butterflies were inescapably drawn to their open and voracious mouthparts. At least, that was the idea. When I went looking for research papers to try to understand how this type of aggressive mimicry worked, I found nothing. No one had ever actually studied why orchid mantises look like a flower, and how they caught their food. This idea had been

---

*   Hingston, J. (1879). *The Australian Abroad: Branches from the main routes round the world*, Sampson Low, Marston, Searle, and Rivington, p. 188.

floating around unchallenged since it was first put forward by none other than Alfred Russel Wallace.

After his humble beginnings selling specimens to fund his research, and the catastrophic loss of his Amazonian specimens in a ship fire, Wallace would go on to greatness. You're no doubt familiar with Wallace as the co-discoverer of evolution by natural selection with Charles Darwin. If Darwin was the Beatles of evolutionary biology, then Wallace was the Velvet Underground. In hindsight, he is revered for being just as influential, but at the time was the punk-rock underdog that never reached the same levels of commercial success. After returning from the Amazon, it only took a few years for Wallace to head off once more in search of adventure and discovery. This time, he headed to South-East Asia and the Malay Archipelago. This is where Wallace made the pioneering observations that would solidify his place in history, and where he first heard tales of these legendary walking flowers.

Like many of the tales of trickery in this book, the idea that the orchid mantis could lure pollinators to their deaths seemed so intuitively obvious that no one thought to question it any further. When I realised that this was a gaping hole in our understanding of mimicry, I knew that it was time to stop staring at tree trunks looking for little brown mantises, and to explore the jungles of Asia looking for bright white mantises. And so, my troubles began; getting lost in Malaysian rainforests, drinking inordinate amounts of Teh Tarik and Pocari Sweat and finding lots of lovely white and pink flowers but very few orchid mantises. I learned why very few people ever attempt studying animals like orchid mantises – they are really tricky to find.

They are incredibly rare animals, and just to make things a little harder, don't look much like animals at all. I couldn't help but gaze around the forest canopy and wonder what else I wasn't seeing, either because I didn't have the physical ability to perceive the sights, sounds, smells and other aspects of the world around me, or because there were endless ways that living things trick, swindle, hustle and dupe other animals. And what about those tigers that still, apparently, live somewhere here in the jungles? In the unlikely event that I ever came near a rare wild tiger, would I have even known about it? Was there ever a tiger watching me while I clumsily stomped through its rainforest home, looking everywhere but seeing nothing?

When I did finally come across live orchid mantises, their impossible beauty lived up to the photographs I had seen online and in books. Their white exoskeleton seemed almost translucent and would shimmer with white glare in full sunlight. What struck me most was how their flowery disguise was achieved by only a few modifications to the stereotypical praying mantis form. If you took a regular green praying mantis, turned it white, then added some petal-shaped expansions to the femurs of its four hindlegs, you would have something that looked just like an orchid mantis, which also makes quite a convincing counterfeit of a white flower.

With a few live orchid mantises in hand, it was time to start testing whether they lured pollinators. The approach was simple: I put live orchid mantises out in the forest, then wait and watch. The results came in much quicker than I expected. When I took a live orchid mantis and placed it on the end of a stick among some shrubby vegetation, it was only a matter of minutes before a small bee flew along and stopped right in front of the orchid mantis. The bee hovered in the air a few centimetres away from the waiting predator. It made a few quick movements side to side, and then flew away again. A few minutes later, another bee did the same thing; it approached the orchid mantis, gave it a close inspection and then flew away. More bees came and each one did the same thing; some took quick glances at the orchid mantis, others took a few seconds to look at it from different angles before darting away. In the shrubs around me were small white flowers where bees were doing the same thing; flying up and facing the flower, hovering side to side, moving closer and closer until they landed on the flower in search of nectar. These same small bees would then fly up to the orchid mantis, hover in front moving side to side, inching closer and closer until – snap! Finally, an orchid mantis lashed out with its raptorial forelimbs. The first few times I saw this, the bees were lucky; the orchid mantis missed, and the bees flew off. Eventually I witnessed, for perhaps the first time in recorded history, an orchid mantis striking with its forelimbs and catching a bee that had been lured in by the mantis's flower-like disguise.

With a few friends and colleagues, I spent weeks watching orchid mantises luring in bees and, occasionally, catching them as prey. We found that they weren't just slightly attractive to bees, they were super-attractive. When we averaged how many bees visited an orchid mantis per hour, they were even more attractive than real flowers growing in the shrubs around them. So, after a century of conjecture, and just like Wallace had predicted, we could confirm that, yes, orchid mantises lure pollinating insects as prey. So, was their status as a classic example of aggressive mimicry justified? Well, as always, it's complicated.

Mimicry, in theory, requires a model; something that animals learn to be either good or bad, which leads them to be fooled by the mimic. In the case of the orchid mantis, there seems to be no model. There is no one species of flower that the orchid mantis has evolved to look like. There are plenty of white and pink flowers growing in the jungles of Southeast Asia, and many that have petals shaped a little like orchid mantis legs, but nothing that stands out as a perfect match for the orchid mantis.

I did another experiment that threw into question the straightforward idea of aggressive floral mimicry. Using some wire and polymer clay, I made some convincing, if not a bit pudgy-looking, models of orchid mantises. When I put them out in the field, bees would fly right up to them just like they would a real orchid mantis. If I changed the colour of the models to brown or green, they were less attractive to the bees, which wasn't a huge surprise. I also changed the model's flower-like appearance by moving around the petal-shaped legs or removing them altogether. Surprisingly, this had no effect on the bees; they were still attracted to the artificial mantises. How then could this be aggressive floral mimicry if the model orchid mantises were still attractive without their petal-shaped legs?

I still don't have an answer to this one. It's possible that the bright white colour of the orchid mantis is enough to attract the attention of bees without the associated petal shapes. In which case, the orchid mantis's trickery may be better described as sensory exploitation rather than aggressive mimicry. Maybe being vaguely flower-like is enough to trick pollinators without being a perfect match to a particular flower. It's also possible that the orchid mantis's petal-shaped legs serve a different function entirely. They could work as a type of masquerade; helping the mantises hide from predators by looking like flowers. There is even recent research showing that the petal-shaped lobes on orchid mantis legs help them glide through the air. When jumping from a height, orchid mantises can steer in mid-air using their legs like miniature aerofoils.

There are still many mysteries surrounding orchid mantises. Like many of the seemingly simple cases of animal trickery in this book, things are rarely as straight-forward as they seem. Similarly, deciding whether the scents of a pitcher plant or the spider-like tail of a viper count as cases of sensory exploitation or aggressive mimicry becomes a hair-splitting exercise that might not result in very convincing answers.

The more we understand about the complexities of deception, the harder it becomes to definitively classify things into neat categories like sensory exploitation or aggressive mimicry, crypsis or masquerade, Müllerian or Batesian mimicry. This goes some way to explaining why sensory ecology, more than many fields of biology, spends a lot of time creating and debating terminology and definitions.

This isn't to say that these concepts aren't useful, or that the ideas of people like Bates and Wallace were wrong – they were revolutionary and accurate as possible for the time. Just like you don't get David Bowie without the Velvet Underground, or 'Purple Rain' without 'Purple Haze,' we don't get a fuller understanding of deception in nature without building from the principles of classical mimicry and camouflage theories.

6

Deception as the key to a successful relationship

Let's take a break from all this talk of death and deception. I wouldn't want this book to detract from your appreciation of nature as a place of serenity, balance and quiet solitude. For just a moment, let's take a walk through a peaceful meadow. Imagine the warmth on your back from the afternoon sun on a spring day. As you walk through the meadow, you reach down with your hands and feel the gentle tickle of grass flowers brushing against your fingers. All around you is lush green grass stippled with small colourful dots, the bright blossoms of this new season. Without a worry in the world, you take the time to sit and watch the comings and goings of buzzing bees zipping through the meadow.

In front of you, a small honeybee makes short meandering flights between petite blossoms. At each blossom, it sticks out its tongue-like mouthparts and sups the rich nectar hidden in the middle of each flower. With each new flower that it visits, the bee's hairy body is dusted with more bright yellow specks of pollen. You start to wonder how the bee perceives each flower, with its multi-faceted compound eyes and photoreceptors that sense a very different palette of colours to your own eyes.

You follow the bee's swirling flight path among the flowers until it leads you towards a small botanical gem: a slipper orchid. These flowers are a rare find. Their pointy striped petals lead your eye towards the centre of the flower, where there is a circular opening to a bulbous lower petal. The bee lands on top of the bulb-shaped petal and meanders towards the opening, searching for what rewards may lie inside. You watch as the bee's entire body disappears into the small hole only to discover that hidden inside this flower, is nothing. The awkwardly cluttered insides of the petal that the bee must navigate force them to squeeze their way out again. Despite the showy petals and tantalising bulb of the slipper orchid's flower, there wasn't a single drop of nectar to be found. There was, however, a curious bundle of pollen stuck to the bee's back as it flew away.

And, just like that, we're back. Even the serenity of a flowering meadow is no refuge from nature's tricksters, and not even the humble flower is innocent of using deception to lure in pollinators. So far, we have mostly focused on animals, but when it comes to using deception for reproduction, plants are the master tricksters. For plants, reproduction often isn't as simple as boy meets girls. Being stuck to the ground means finding some way for your male sex cells (i.e. pollen) to get to the female sex cells inside another plant. In flowering plants, the solution involves a complicated *pas de trois*, which is French for ... I don't know, a type of pastry or something? Anyway the point is that there are three parties involved – the male plant, the female plant and an animal pollinator which brings pollen from one to the other. This relationship is usually maintained by flowers offering their pollinators a reward in the form of nutritious nectar. However, many pollinators fall victim to foul play as numerous plant species, like slipper orchids, don't provide them with any reward whatsoever.

Of all the varied types of flowing plants, the most successful tricksters are orchids. We prize them for their beauty and allure, but as far as pollinating insects should be concerned, orchids are not to be trusted. They are among the most speciose flowering plants in the world, with current estimates sitting at almost 30 000 named orchid species. Almost a third of all known orchid species are thought to be deceptive. The flamboyant blossoms of slipper orchids, vanilla orchids, marsh orchids and spider orchids, to name a few, make nothing but false promises.

Showy blossoms are often enticing enough to lure in curious insects that brush past bundles of pollen on their way to discovering how overpromising these flowers truly are. Other orchids have refined this tactic and have evolved to closely mimic nearby rewarding flower species. The South African orchid *Disa nervosa* is a remarkable mimic of a completely different flower, the iris *Watsonia densiflora*. Both plants' inflorescences are made of tall stalks of bright pink flowers. The two species' flowers are the same size and have a similar funnel-like shape. And they are both pollinated by the same type of insect, a long-tongued fly, *Philoliche aethiopica*. Despite being two completely unrelated species, the plants are remarkably similar but for one glaring difference: the iris provides a nectar food reward for the flies, and the orchid doesn't. Flies that learn to associate the pink flowers of the iris with food can be easily tricked into the visually and chemically similar but inevitably unrewarding orchid. Similarly, the European globe orchid *Traunsteinera globosa* looks barely like an orchid at all. Instead of the large showy flowers we generally picture when we imagine an orchid, their flowers are tiny and are clustered together in a dense circular bundle. At a glance, they look like the small flower clusters of herbs rather than orchids. Research has shown that not only do the orchids match the colours of co-occurring perennial herb species, but also give off similar scents. Again, flies duped by these similarities will readily visit these orchids despite their not having any nectar. Just as mimicry can be used for defence or to lure in prey, mimicry is also used by plants to lure in pollinators.

It's easy to imagine how pollinators would be tricked into visiting these bright colours with the promise of sweet nectar. But the false promises presented by some other plant species are a little harder for us to relate to and require us to adopt the perspective of their animal visitors. The dead horse arum (*Helicodiceros muscivorus*) has one of the most matter-of-fact common names in botany; it is a species of *Arum* that smells like a dead horse. Combine this scent with its large pink, fleshy, warm, hairy inflorescence, and the dead horse arum appears less like a blossoming flower and more like a rotting carcass. While this might not sound particularly appetising to you, it smells of sweet nectar to carrion flies that feed on putrid animal corpses. Flies drawn in by the horrid smell land on the blossom and walk towards a warm, hairy hole at the base, which some have suggested resembles an open, hairy rectum. From pictures I have seen of the plant, that description isn't far off. Once the hungry flies walk into the false bunghole, they are trapped there overnight by the thick, pointy butt hairs. When morning comes, the false butt hairs wilt and the still-hungry carrion flies can escape, collecting more pollen from the plant on the way out. Several closely related *Arum* species all use a similar tactic. Combining putrid scents with their unfortunate appearance and generating small amounts of heat at the centre of the flower, they manage to attract a suite of carrion- and dung-feeding insects as pollinators.

This deceptive technique isn't restricted to arum plants. The two largest flowering structures in the world are also famous for their putrid stench that lures in flies and beetles that feed on dung and rotting organic matter. The plant with the world's largest inflorescence of flowers has the species name *Amorphophallus titanum*, which roughly translates to 'giant misshapen penis'. Its common names are arguably more tasteful – the corpse flower or titan arum. Native to Indonesia, they are extremely rare plants, and the blossoming of a corpse flower is an attraction that draws tourists to botanic gardens in the thousands across the globe. The flower first emerges from the soil as a large bullet-shaped bud. Then, over about a month, the bud grows straight upwards, eventually unfurling a single enormous leaf to reveal the mighty inflorescence* – a yellowish wobbly 'phallus' that stinks of rotting flesh. The largest corpse flowers ever recorded reached over 3 metres in height and

---

*   The individual flowers of *Amorphophalus* plants are very small. The inflorescence describes the enormous structure where the tiny flowers are clustered and so tightly packed that they are often described as a single flower, when it's actually a clustering of many small flowers.

weighed over 100 kilograms. The blossoming only lasts for about two days, during which they are visited by swarms of flies and beetles. After two days, the plant loses its foul odour, and the entire flower withers and decays.

The plant with the largest single flower also comes with a prodigious stench. This attracts carrion flies not just as a false promise of food, but also as a place to lay their eggs. Rafflesia flowers grow on the ground and the largest species (e.g. Rafflesia arnoldii from Sumatra and Borneo) have flowers that can reach over a metre in diameter. They have leathery reddish-brown lobes that surround an enormous bowl-shaped opening that is big enough to fit your head into. Believe me, I've tried. The foul odours that emerge from this opening lure in flies that would usually lay their eggs inside dung or rotting carcasses. This tactic, called 'brood-site mimicry', is used by many other deceptive plants. The aptly named skunk cabbage (*Symplocarpus foetidus*) and carrion flowers (*Stapelia* spp.), as well as the dutchman's pipe (*Aristolochia gigantea*), Indian almond (*Sterculia foetida*) and several thousand other plant species appear to use foul odours and brood-site mimicry to attract pollinators.

The Chinese orchid *Dendrobium sinense* is pollinated by a predatory hornet, *Vespa bicolor*. The orchid releases chemicals that mimic honeybee alarm pheromones, thus attracting hornets on the hunt for bees. Similarly, the Middle Eastern orchid *Epipactis veratrifolia* smells like aphid alarm pheromones and attracts predatory hoverflies that specialise in feeding on aphids. These and many other bizarre pollination strategies seriously challenge our perceptions of flowers as mere pretty and delicately perfumed ornaments.

Deceptive pollination has puzzled biologists for a long time. Firstly, because it seemed so unnecessary. The mutually beneficial partnership between nectar-producing flowers and pollinating animals makes perfect sense. Why would a plant jeopardise that relationship? To save a few calories on nectar production? Probably not. There is more in it for the plant, and deceptive pollination is actually a clever way for plants to avoid inbreeding. Pollination is a great system for plants to get pollen from one plant to another, but it's not perfect. Insects visiting nectar-rich flowers are likely to come back to the same flower, or nearby flowers on the same plant, looking for more food. This increases the likelihood of pollen being brought straight back to the same plant, resulting in self-fertilisation. When pollinators visit a flower only to discover that is has no nectar, they tend to abandon that patch of flowers quickly and fly away to a different place in search of more profitable feeding grounds. The pollen they picked up on the deceptive flowers is thus more likely to end up on a different individual.

The pollinators might start to get a bit suspicious if they visit too many unrewarding flowers, but it's still a worthwhile strategy for the plants. Since the only cost to the pollinator is a little lost time and energy, the incentives for them to learn and avoid specific flowers are rather negligible and deceptive flowers can still expect a steady stream of curious pollinators coming to visit. At least, it was for a long time assumed that the costs for pollinators were low. Certain deceptive flowers have more intricate ways of tricking their pollinators that can leave them missing out on more than just a meal.

Sexually deceptive pollination

The unlucky bees that stumble across *Ophrys* orchids may initially be drawn towards the flowers' bright triangular petals.* As they approach, however, other aspects of the flower draw their attention. Quickly the bees' interest in foraging disappears. Hanging from the base of this flower is a strange round mass. It has dark banded patterns and stands out against the flowery colours of the petals behind. The purpose of this strange structure, called a perigon, becomes clear when the bee lands on it, wraps its legs around either side and vigorously attempts to mate with it. In the process, the bees press their heads against a special pair of pollen bundles that get stuck to their hairy bodies and are carried away to the next flower. *Ophrys* orchids release chemicals that mimic the scent of female bees. Together with their strangely shaped perigon, the orchid appears to convince male bees that the flower is a female bee. Of all forms of floral deception, perhaps the most elaborate belong to the thousands of plant species that dupe male insects into attempting to copulate with flowers.

From a human perspective, there often isn't much to appreciate about the similarity between sexually deceptive orchids and female insects. Some have a vaguely insect-shaped bump, or dark colour patch on their petals. Beyond this, they don't look like insects at all; it takes large amounts of imagination to understand what male insects could see in these temptresses. Flying duck orchids

---

* For the botany boffins reading: yes, I keep using the word petals to describe what are technically sepals. Please accept my humblest apologies.

(*Caleana* spp. and *Paracaleana* spp.) get their name because they look weirdly like duck heads. The duck heads have an important purpose, which we'll touch on soon, but it doesn't have anything to do with their similarity to ducks. Tongue orchids (*Cryptostylis* spp.) could be likened to drooping tongues, if those tongues were suffering from some heinous disease that inflicts them with dark purple warts, green stripes and red fur. Greenhood orchids (*Pterostylis* spp.) look like inconspicuous curled green leaves that many people might walk past not even realising that they are an orchid. Despite appearances, each one of these orchids manages to convince male insects to try their hand at knocking boots with a flower.

Again, when we adopt the right perspective, it makes perfect sense. To male wasps searching for females, good looks pale in significance to the right scent. Sexually deceptive orchids primarily dupe their pollinators using aromas that mimic female wasp pheromones. They are so effective that when insects attempt to mate with the flower, they don't just make a fleeting visit and quickly realise their mistake. Males will enthusiastically try to mate with the flower and can be coaxed into carrying out elaborate behaviours usually reserved for females of their species. Some studies have shown that male insects will preferentially choose to mate with flowers over real females, or abandon a real female while mating to instead have a crack at a nearby flower. *Cryptostylis* tongue orchids lure in male wasps that land on the flower, then turn upside down and walk backwards, inserting their abdomen into the base of the flower. The orchids' pollen packets then get stuck to wasps' abdomens, where they are perfectly placed to transfer pollen to the next orchid that they poke their rear end into.

Hammer orchids (*Drakea* spp.), and elbow orchids (*Arthrochilus* spp.) use their pollinators' mating behaviour against them. Wasps that visit these elaborate flowers don't just get fooled, they get a solid beating at the same time. These orchids are pollinated by thynnine wasps. When these wasps mate, the male finds a female, pounces on her back, grabs her tightly with his legs and flies away with her, where they mate 'on the wing'. When a thynnine wasp is tricked into mating with a warty hammer orchid, they are lured towards an elaborate scoop-shaped structure covered in round bumps that sits off to one side of the flower on a thin bent stem. Mistaking this warty mass for a female, the male wasp grabs on tightly and tries to fly away. However, the thin stem connecting this structure to the flower is hinged, and when the wasp tries to fly away, he is quickly swung around in a tight arc and he is quite literally hammered head first into the other side of the flower where there are tightly packed bundles of pollen. Flying duck orchids use a similar trick. What could be described as the duck's head sits high above the flower on the end of a curved stem (the duck's neck?). This curved stem acts like a spring-loaded trap, and is kept extended by a build-up of fluid pressure inside the plant. When a wasp tries to grab the duck's head, the fluid pressure is released and the wasp is flung upside down into the undersides of the flower.

In 2008, scientist Anne Gaskett made a surprising discovery when studying *Cryptostylis* tongue orchids. They are pollinated exclusively by male *Lissopimpla excelsa* wasps, more commonly called 'orchid dupe wasps'. They vigorously attempt to mate with the orchid in a behaviour that's commonly called 'pseudocopulation'. But Gaskett and her colleagues discovered that there was nothing pseudo about it. Male wasps 'finish the job', so to speak, and ejaculate into the flower. And not just a little bit of ejaculate either; further studies showed that wasps could waste around 10 per cent of their total stored sperm on a deceptive orchid.

This research showed clearly that, when tricked by a sexually deceptive orchid, pollinators stand to lose much more than time, energy and a bit of dignity. And it doesn't end there. *Cryptostylis* orchids don't just play around with *individual* male wasps. In the long term, these orchids have the power to manipulate populations of wasps across generations. It all hinges on the biology of the wasps and what determines whether a wasp is male or female, which turns out to be a little bit complicated.

In many animals, sex is determined by the combination of sex chromosomes in that animal's DNA. For most mammals, a pair of X chromosomes will lead to the development of female characteristics, whereas one X and one Y chromosome in the genome leads to the development of male characteristics. Many birds use a similar system in reverse; males have matching chromosomal pairs (ZZ) and females have a pair of different chromosomes (ZW). Other types of animals have completely different mechanisms that determine sex; crocodiles, turtles and a few bird species have their sex determined by the temperature at which the eggs develop. And in certain insects, like ants, bees and wasps, sex is determined by whether an egg is fertilised or not.

After mating, a female wasp can store male sperm and use it to fertilise her eggs. She can also lay eggs without using male sperm. These eggs will still develop into viable offspring. The wasps that come from fertilised eggs develop into females, and those that come from unfertilised eggs develop into males. The technical term for this is *haplodiploidy*. If an egg is fertilised, the female that develops is diploid, meaning that it has two sets of DNA in its cells – one inherited from the mother, one from the father. If an egg is unfertilised the male that develops is haploid; it only has one set of DNA, inherited from the mother.

Since *Cryptostylis* orchids are so overwhelmingly attractive to male wasps, and are very good at persuading them to part with their sperm, it's possible that female wasps are missing out on potential mates and their sperm. Since the sex of wasp offspring is determined by whether eggs are fertilised or not, would that mean that dupe wasps are less likely to lay female eggs and more likely to lay male eggs? This is precisely the question asked by scientist Amy Brunton Martin, along with Anne Gaskett, from the University of Auckland. They scoured museum records of these wasps across eastern Australia and found a surprising pattern. In areas where there are no deceptive orchids around, wasp populations were roughly 40 per cent male. But in areas where there are sexually deceptive orchids, the sex ratios are skewed and males make up over 70 per cent of the wasp population. So, by tricking male wasps into copulating with their flowers, these orchids don't just get a handy pollination service, they also reduce the chances that the male mates with a female wasp. Therefore, those female wasps are more likely to produce more male offspring. Which is great news for the orchids because, in the next generation, there should be more males around to dupe into pollinating orchid flowers.

For a long time, scientists thought that deceptive flowers had a benign effect on their pollinators, and that a little lost time spent getting jiggy with a flower was all a bit of harmless fun. The discovery that male wasps could get so carried away with this doting detour that they could waste entire sperm packets on a flower seriously challenged this assumption. Then, the discovery that orchids could actually manipulate the sex ratios of their pollinators showed once and for all that deceptive flowers can have enormous impacts on their pollinators. But let's be reasonable, some lost time here and spilled sperm there is still no big deal, right? It's not like deceptive flowers are getting away with murder … right?

The bright light at the end of the pollen tube

There are always those creatures that have to take things a little bit too far. In the world of deceptive pollination, those scumbags are *Arisaema* flowers. Sometimes called cobra lilies or 'jack-in-the-pulpit', they are cultivated as ornamental plants for gardens and have stunning blossoms, which might go some way to account for their horrendous personalities. Their ornamental 'flowers' are more accurately termed a spathe – a modified leaf wrapped into a vertical tube with a large hood that folds over the top. In the centre of this tube is a thick protruding stalk, which is a dense cluster of flowers so small they are hard to see with the naked eye. Imagine a peace lily, but more sinister-looking, and you're close to picturing an *Arisaema* blossom.

They are mainly pollinated by flies and fungus gnats – small flying insects that lay their eggs on fungi. Depending on the species, the flowers can use either brood-site mimicry or sexual deception to lure in pollinators. Female insects are drawn to the mushroomy smells of certain *Arisaema* flowers when searching for somewhere to lay their eggs. Other *Arisaema* species seem to mostly attract male insects, suggesting that they may be lured by false pheromone cues.

What happens to the bamboozled pollinators when they land on one of these flowers depends on whether it is a male flower or a female flower. When flies land on a male flower, they are drawn to the central cluster of flowers and wander about its surface getting thoroughly coated in pollen grains. As they explore deeper into the tube-like spathe, the passageway narrows. A slippery inner surface sends gnats tumbling down into the base of the tube, where downward-pointing hairs stop them from climbing back out again. There is only one direction for the gnat to go: down. Eventually, they reach the base of the spathe, where a convenient gnat-sized hole allows them to escape.

When flies land on a female flower, the process is similar except for that crucial escape hole. Upon landing on the female inflorescence, presumably transferring pollen in the process, the flies may still walk or fall towards the base of the spathe, where the downwards-pointing hairs create a one-way path to nowhere. Eventually they reach the bottom, where they are trapped for good. Male flowers let their pollinators escape to deliver the pollen elsewhere, whereas female flowers have evolved a deadly solution to keeping that pollen, and the pollinator, all to themselves.

It is still unclear what benefits *Arisaema* flowers gain from trapping their pollinators in a floral tomb, and why insects haven't learned to avoid them, given the risk of death. Recent research has found, however, that death might not be the end for some of these fungus gnats and some *Arisaema* species might not be as bad as they seem. A team of scientists from Japan studied the species *A. thunbergii*, which lures in female fungus gnats with scents that resemble those of fungi. The reason fungus gnats lay their eggs on fungi is that when the eggs hatch, the larvae emerge in close proximity to a nutritious and readily available food source. When fungus gnats are drawn towards *Arisaema* flowers there is no nutritious site for the fungus gnats to lay their eggs. Or so it was assumed.

Looking closely at the thick inflorescences of female *A. thunbergii* flowers, the Japanese scientists found that the gnats were laying eggs and wedging them in the gaps between flowers, and when the eggs finally hatched, the larvae started eating the now decomposing flowers. So, even if the gnats wandered into the base of the spathe, they may have had the opportunity to lay eggs before becoming trapped. And the larvae that hatched from those eggs feed on the flowers, giving them a chance to live on after the demise of their dear mother. This raises the possibility that in this species, pollination isn't entirely deceptive. Perhaps among the hordes of lethal *Arisaema* flowers, there are an honest few that provide a genuine reward to their pollinators in the form of a nourishing nursery for their eggs, in exchange for mum paying the ultimate price.

How animals deceive their way to the devil's dance

It's often said that to succeed, an animal must do the 'three F's': feed, flee and, you know, bump uglies. In previous chapters, we covered how deception is used to do the first two: score an easy meal and avoid becoming a meal. While plants have mastered using trickery for the third 'F' by deceiving pollinators, animals aren't exactly prudish about pulling tricks to pull a mate.

As anyone who has ever *wooed a lady* knows, the first thing you must do is never refer to it as *wooing a lady*. The second is to bring food. And if you can't do that, fake it till you make it. Male nursery web spiders (*Pisaura mirabilis*) impress females by bringing them a gift in the form of a delicious dead insect. Rather romantically, they gift wrap it in a thin layer of silk. If the female accepts, the male may be allowed to mate with her. Rather unromantically, there are sneaky males that will trick a female by gifting her bits of bark and leaves, or the empty exoskeletons of a dead insect concealed in thick layers of silk. Giving females food also helps the male spider by keeping the mouthparts of the often sexually cannibalistic female busy. Just in case the female is in a cannibalistic mood, the male spiders add a secondary bluff to their courtship – they play dead. Once the female has accepted the gift, the male suddenly seems to drop dead and wait for the female to start eating her silk-wrapped meal. Once her attention is on food, the male tentatively comes back to life, creeps in underneath the female, and gets down to business.

In another animal, it's the females that play dead to trick males. Except they don't do it to try to get a mate, they do it to avoid mating. Female moorland hawker dragonflies (*Aeshna juncea*) are approached by males while they are flying. If one is getting pestered by a randy male she's not interested in, she can drop out of the sky, crash-land on her back and lie motionless. After a few moments of bewilderment, the male flies away and the female jolts back to life.

Compared to other forms of deception in the animal kingdom, cases of males duping females into mating are relatively rare. Females are pretty choosy in who they decide to mate with. This means that male animals tend to go to great lengths to display their prowess and quality with bright plumage, elaborate antlers and many other elaborate signals. Faking these signals is tricky and any males who try are quickly weeded out of the dating pool by picky females. Certain freshwater fish manage to trick females into reproducing and they do it in a counterintuitive way; they have specialised fins that look like fish eggs. This works because of the peculiarities of these fishes' mating systems. In striped darters (*Etheostoma* spp.), for example, the males do the important job of looking after the eggs. Males tend to nests where they protect eggs from predators, parasites and rival males until they hatch. When a female selects a male, she will lay eggs in his nest site, then the male will fertilise them by depositing his sperm on top. One very sensible cue that females use to pick a responsible male is to find one that already has eggs in his nest. This is a direct indication of that male's ability to successfully protect and maintain a nest. This poses a conundrum for males just starting out – how do they get that first batch of eggs? The solution is a good old-fashioned swindle. Striped darter males develop curious egg-shaped blobs on their fins. This has been described as a form of egg mimicry and is clearly convincing enough to tickle the fancy of females who are drawn towards the cluster of egg-shaped fin bumps before laying eggs in an empty nest.

One other counterintuitive trick that male animals use to play the mating game is to disguise themselves as female. In rove beetles, some males grow to be big brutes with large mandibles for fighting, and other males are smaller and near indistinguishable from females. Similarly, flat lizards (*Platysaurus broadleyi*) have males that develop the black and white stripes of females rather than the bright red and blue colouration of other males. Some swordtail fish males develop dark patches on their underbellies that look like the dark swollen ovaries of female fish. These gender-flipping tactics work not by fooling females, but by tricking rival males.

Let's take a look at ruffs (*Calidris pugnax*), for example. In these shorebirds, males and females look very different. Females have feathers delicately banded in brown, black and white. Males are larger, with an enormous fluffy crest around their necks and heads that fans out during their mating display. But scientists have discovered that a small portion of males never develop the classic male plumage. Instead, they look just like females. They are relatively small and have the same banded feather patterns. Despite appearances, they are unambiguously male, and have the balls to prove it – female-mimicking males have testes that are over double the size of regular males. During the breeding season, the run-of-the-mill large male ruffs gather in coastal wetlands and spend much of their time showing off their elaborate plumage, puffing their chests around other males and getting into heated fights with their rivals. Meanwhile, the smaller female-mimicking males appear to go unnoticed. They can wander through the wetlands in search of females, bypassing the ritualistic competitions of other males and putting their prodigious testicles to good use.

Early on in the book, we talked about the spectacular ability of cuttlefish to change their colour. This superpower comes in handy when trying to weasel your way towards a female. The courtship behaviour of Australian giant cuttlefish (*Ascarosepion apama*) is a spectacular event where males show off using their technicolour skin displays. As they sidle up to females, male cuttlefish display dark circular bands that flow over their backs like waves and fan out their broad arms that shine bright white along the edges. During courtship, the females adopt a mottled brown and white pattern. If any rival males approach a courting pair, the displaying male must interrupt his performance to chase away his rival and sometimes engage in brutal combat.

Just like in the examples above, this is a challenge for smaller males who can't match the fighting abilities of larger males. But it is a challenge easily overcome with some clever colour-change tricks. A team of scientists observed the mating behaviour of giant cuttlefish in the wild off the coast of South Australia and found that courting pairs of cuttlefish were often closely followed by a third wheel – a much smaller male. These males seemed to avoid the aggression of the much larger males by adopting the mottled brown and white colour pattern of a female cuttlefish. They would wait patiently for another large rival male to turn up. Then, when the larger males were distracted chasing away rivals, the small males would quickly move in towards the female, switch colours to the male courtship display, and dart in for a quickie with the female, while the bigger bloke was distracted.

Another species of Australian cuttlefish, the mourning cuttlefish (*A. plangon*), has refined this trick even further. Since they have such precise control over their colour patterns, mourning cuttlefish go so far as to display different patterns on different sides of their bodies depending on who is watching. On the side of their body facing a female, a male can display bold courtship zebra-stripes, while simultaneously disguising themselves as a female to any nearby males by showing mottled brown patterns on the other side of their body. These patterns can be rapidly switched and flipped depending on whether males or females are present and in what direction they are.

This sort of complex behaviour starts to raise some thrilling questions about cuttlefish minds. Are they intentionally lying to each other? Do they understand concepts such as honesty and dishonesty? Can they adopt the perspective of another animal to effectively use deceptive behaviour against them? A short one-word answer to these questions could be 'yes'. A more accurate two-word answer is 'it's complicated'. Whether deception is ever intentional in nature is a complex topic. The same could be said of deception in humans. But that is a conversation I will attempt in a later chapter.

From sexual deception to sexual parasitism

As icky as all this sexual coercion sounds, I don't want to give males a bad rap. Perhaps the most infamous of sexual deceivers are females that trick males into handing over their sperm without any intention of ever using it. What makes this story even weirder is that these females don't even trick males of their own species, they trick males of completely different species into mating with them. It sounds strange, doesn't it? That's because it is. Such is the weird biology of the Amazon molly (*Poecilia formosa*), a freshwater fish from North America that seems to break the rules of every biology textbook.

On first impressions, the Amazon molly looks like your everyday fish. They are small guppy-looking things, that swim about, flap their fins and do all the regular things fishes do. The females give birth to live young, like many other freshwater fish do, and to make those babies, the female Amazon molly does what most other females do – finds a male, does the deed, and uses his sperm. Bada-bing bada-boom, baby fishes. There's only one problem, and it's a doozy of a pickle; there's no such thing as a *male* Amazon molly.

Amazon mollies as a species are entirely female. Yep, every single one of them. And the babies they give birth to are clones of the mother. If you look at a family tree of freshwater mollies, there are two species of fish closely related to Amazon mollies that have both males and females: the sailfin molly (*P. latipinna*) and the Atlantic molly (*P. mexicana*). Genetic analyses suggest that the Amazon molly species arose from sailfin mollies and Atlantic mollies interbreeding, leading to a third, hybrid species. Hybrid species are rare in nature, but they can happen, and in this case the end result was an entirely female clonal species. While the Amazon molly's clonal abilities sound impressive, evolution often lands on imperfect solutions and this all-female species didn't develop a way to make babies without using male sperm as part of the process.

Sperm doesn't just provide genetic material for making offspring. The action of a sperm cell interacting with an egg cell triggers biochemical reactions that kickstart the growth and development of that egg into an embryo. This is a process called embryogenesis. So, while Amazon mollies don't need males to provide genes for their offspring, they still need sperm to kickstart the embryogenesis process. So, what's a girl to do? Find the next best thing – sperm from a different species of fish.

Given their hybrid origins, Amazon mollies are biologically very similar to both sailfin mollies and Atlantic mollies. Which is a lifeline for the Amazon mollies because the sperm of both these species is suitable enough to penetrate the Amazon mollies' eggs and kickstart embryogenesis. Sperm from a fourth species of molly, the Tamesi molly (*P. latipunctata*), will also do the trick. Research has shown that male mollies are entirely capable of distinguishing between females of their own species and Amazon mollies. But it also shows that they are often not choosy about who they decide to mate with and can be duped into some interspecies yentzing with Amazon mollies.

After mating, the genetic material inside the sperm is discarded and the embryo develops as a clone of the Amazon molly mother. Because the genetic material inside the sperm is effectively wasted, Amazon mollies are often called *sexual parasites*. Evolutionarily speaking, that sperm would be better spent fertilising eggs of their own species. Instead, it is coaxed from them by genetically incompatible Amazon mollies. How this strange reproductive system evolved, and why it persists in Amazon mollies, continues to fascinate scientists. It's a roundabout solution to an unnecessary problem, but such is the imperfect nature of evolution.

We've covered some of the many ways that both plants and animals can use trickery to achieve the three F's.* And as important as these three things are, animal life cycles are complex and there are many other things that creatures need to achieve throughout their lives. Deception, it turns out, is a lifelong strategy for some animals, and there is one group of animals that are perhaps the most famous tricksters in the animal kingdom. So much so that other deceptive animals are named after them in their honour. They're such infamous tricksters that many of you reading might have expected to meet them earlier in the book. They are the notorious cuckoos.

---

* Which, in case you needed a reminder, are: feeding, fleeing and getting one's kettle mended.

7

Deception:
a life-long calling

Understanding nature can help you in your daily life. And I don't just mean life hacks like keeping a compost bin or growing your own parsley. I've found that viewing the world through the lens of evolutionary biology has given me surprising moments of solace in challenging situations. Parenting, as many of you know, is one of the hardest challenges a person can face. It's incredible how, once you are a parent, the particular tone and timbre of your own child's scream pierces into your mind. It fills you with a sudden and crippling anxiety that only you can understand. I was helped immensely as a new parent in these moments when I realised that the screaming pink squishy thing in my arms wasn't necessarily a child in distress, it was simply a small animal making small animal noises. It was a fledgling chick chirping loudly in its nest because that's what chicks in nests do. My child was squawking needily because that's what babies have evolved to do. And my brain was triggered to respond anxiously to these squawks because that's what it has evolved to do. Just as the gentle chirp of a chick in its nest triggers certain responses in their bird parents, guttural screams from a human baby are there to trigger behaviours in their human parents.

That sense of crippling anxiety that I felt in response to their screams wasn't a sign that there was something wrong, or that I was failing in my duties. It was there to trigger me into doing something and, just as a daddy bird goes looking for grubs and worms to shove into the gaping mouths of chicks, I felt compelled to microwave bottles, change nappies, tighten swaddles or whatever the hell it took to get the thing to shut its mouth. This mindset helped me greatly in these moments as it turned the screaming-baby stage of parenting from a stressful experience into a run-of-the-mill biological process that I, like many other animals, was obliged to go through.

Having said that, this only worked some of the time. Most other times, parenting continues to be the draining uphill battle we all know that it can be. Like early this morning, for example: I sit typing this paragraph, dreary and grumpy thanks to my two-year-old who thought that 3 am was a great time to wake up and tell me all about the baby lizard she saw on a rock the day before. At least, I *think* that's what it was – it was hard to decipher toddler speak with a sleep-deprived brain. I'm sure that one day I'll look back on that moment as an adorable and funny story. At 3 am this morning, however, I was less optimistic – I began to wonder, as every parent does at some stage, why me? Why do I have to be the one here lying groggily on the couch begging a toddler to shut the hell up? Surely there's a way around this. There must be a way to fulfill this biological imperative to pass your genes onto the next generation without having to deal with all this parenting garbage. Fellow parents, let me introduce you to a concept called *brood-parasitism* – nature's solution for getting other parents to do all the hard work for you.

The cliff swallow (*Petrochelidon pyrrhonota*), for example, invests significant time and energy into sculpting large domed nests from mud pellets that hang from the undersides of rocky outcrops. Upon laying the eggs, the cliff swallow must incubate the eggs. Then, after the eggs hatch, they must spend significant amounts of time and energy balancing the priorities of protecting the nest and collecting food to bring back and feed the hatchlings. Every now and again, a cliff swallow seems to decide that she's just not in the mood for all that malarkey and goes and lays an egg in someone else's nest. The adoptive parents seem to not notice one extra egg in the nest and will continue protecting and rearing another bird's offspring as if it were their own. The parasitic cliff swallow exploits the parenting instincts of its neighbours and can pass on its genes without having to invest in rearing its own offspring. This strange behaviour, where a bird lays its eggs in the nest of another bird of the same species, has been seen in a few hundred different bird species.

Despite their widespread occurrence, cases of brood parasitism are opportunistic and relatively rare. It is still the norm for birds to build their own nests and feed their own chicks. But there are other birds that have refined the art of brood parasitism into a complex lifestyle. Cuckoos (family Cuculidae) are notorious for parasitising other birds' nests. Across the globe there are around 150 species of cuckoos and about 40 per cent of all cuckoos are called 'obligate brood-parasites' – they never build their own nests, they never incubate their own eggs and they never feed their own nestlings. The only way that they can breed is by laying their eggs in other birds' nests. They do this by using a suite of adaptations that make them masterful manipulators of other birds' parenting instincts. For these birds, deception is a lifelong vocation. Their biology from egg through to adult requires that they trick, leech on and bully other birds.

The most-studied cuckoo is the common cuckoo (*Cuculus canorus*). It lays its eggs in the nests of reed warblers, dunnocks, robins and several other types of birds. The first challenge for the common cuckoo is to get access to the nests. The whole point of nesting in the first place is so that breeding birds can protect their eggs, and parents will viciously try to fight off potential nest predators and parasites, including cuckoos. Despite their small size, nesting birds like reed warblers will swoop and peck at larger birds. By making loud alarm calls, they can alert other nearby nesting birds to the presence of a threat and attract a mob of defensive parents to fight off the intruder. Though there is a limit to what small nesting birds can put up a fight against. Small reed warblers are no match for large birds of prey, and they are unlikely to attempt mobbing something as dangerous as, say, a hawk.

This is where the cuckoos' first deceptive trick comes into play. Some cuckoos look remarkably like birds of prey. The common cuckoo has been likened to the sparrowhawk (*Accipiter nisus*), as they both have bright yellow feet and eyes and bold black and white striped plumage on their underside. Female cuckoos also have calls that sound a lot like sparrowhawk calls. Other parasitic cuckoos have passing similarities to harrier-hawks, goshawks and black bazas. The African cuckoo-hawk (*Aviceda cuculoides*) is so named because of its noticeable similarity to common cuckoos. Striking raptor-like plumage isn't seen in non-parasitic cuckoos and this has led scientists to wonder whether this was some form of mimicry. Cuckoos could use their frightening appearance to intimidate smaller birds that would otherwise mob a parasitic cuckoo.

One study tested this by taking taxidermied mounts of common cuckoos and placing them near reed warbler nests to see whether they would be attacked or not. When the researchers covered the black and white barred plumage on the underside of the cuckoos with white fabric, they were much more likely to be attacked by reed warblers than when their stripes were visible. They repeated the experiment with stuffed sparrowhawk specimens and saw the same pattern: the birds with barred plumage were attacked less frequently than the ones where the stripes weren't visible. The conspicuous banding on cuckoos seemed to make them more intimidating to reed warblers, probably because of the similarity of the plumage to that of a sparrowhawk. Other studies presented modified cuckoo mounts near bird feeders and found that striped cuckoo mounts were intimidating enough to frighten smaller birds away from the feeder, whereas mounts without stripes were less intimidating.

Even if a cuckoo's intimidating plumage isn't enough to frighten birds away from protecting their nests, it is sufficient to make other birds think twice about mobbing the cuckoo, on the slim chance that it really is a hawk. That can allow the cuckoo to safely sit back and play the waiting game; patiently watching until the guarding parents fly off in search of food and the cuckoo can fly in to take over the nest.

Once a cuckoo has gained access to a nest, it will quickly lay an egg in with the other bird's brood. They usually only deposit a single egg – this presumably helps the egg go unnoticed by the parent when they return to the nest. If their clutch were to double in size, the nest owners might start to suspect something is awry. One extra egg in a clutch is much less likely to arouse suspicion. Some species, such as Horsfield's bronze cuckoos (*Chalcites basalis*), cleverly remove a single egg from the nests of their hosts before laying their own. It couldn't be that simple though, could it? This isn't like a cliff swallow laying another cliff swallow egg into a nest already full of cliff swallow eggs. This is a bird of one species laying an egg into a nest of a completely different species. Surely an egg from a different species would stand out against the original clutch of eggs? This is where the cuckoos next deceptive trick comes into play.

Different bird species' eggs look remarkably different and can have speckles and markings on the shell that are characteristic for each different species. Cuckoos have evolved eggs that match the colour patterns of their host's eggs. As mentioned above, common cuckoos don't just parasitise a single species of host bird, they parasitise several different types of birds, each with different types of eggs. However, within the common cuckoo species are several distinct races that specialise in parasitising different types of birds. For example, cuckoos that parasitise black-faced bunting nests lay white eggs with reddish speckles, just like black-faced bunting eggs, whereas those that parasitise marsh warbler nests lay eggs with black and grey splotches, just like marsh warblers. This one species breeds across most of Europe and Asia and migrates to overwinter in sub-Saharan Africa. The evolution of different races within the one species means that the common cuckoo can parasitise different host birds across its entire intercontinental distribution. Similarly, the Australian pallid cuckoo (*Heteroscenes pallidus*) has four different host species and across its range lays different egg types that match the colour of the local host species' eggs.

Not all cuckoo eggs match the appearance of their hosts, and some host birds are better than others at detecting when an unwanted egg has entered their nest and can either discard or destroy the invading cuckoo egg. But if all goes to plan for the cuckoos and their eggs go unnoticed, the host parents will incubate and protect the cuckoo eggs along with their own right up until they hatch. And this is where the real drama starts. Cuckoo hatchlings are more than just sneaky tricksters, they're savage brutes. Common cuckoo eggs tend to hatch earlier than the eggs of their host. This gives them a head start on their nestmates, but they are not simply content to be the early bird that gets the worm, they are instinctively driven to make sure they are the *only* bird getting the worm. As soon as they have hatched, and while they are still blind and featherless, cuckoo chicks start forcefully evicting the other eggs from the nest. Waddling clumsily and blindly around the nest, one by one the cuckoo chick backs up to each egg and, pushing backwards and upwards with its legs, scoops up the egg and pushes it over the side of the nest, letting it drop to the ground below. They have even evolved a noticeable concave curve to their rear ends, which turns their entire tail end into a specialised scoop, perfect for getting under and lifting other eggs out of the nest. If, by chance, the cuckoo chick isn't the first egg to hatch or if their nestmates hatch before their eggs have been evicted, it doesn't make a difference; the cuckoos are just as adept at scooping up live and squirming hatchlings and booting them out. Many other species of cuckoos show similar behaviours whereby they forcibly evict their nestmates. Striped cuckoos (*Tapera naevia*) take a more thuggish approach and have specially adapted needle-sharp bill hooks that they use to viciously attack and kill their nestmates. With their adopted siblings now out of the picture, cuckoos have sole access to the host parent, and their services of food and protection.

You would think that at this point, when there is an invasive bird of a completely different species sitting in their nest, the parents would surely realise that something is amiss with their offspring. But cuckoos have another list of deceptive tricks to convince their new parents to care for them as if they were a child of their own. In some cases, cuckoo chicks seem to do this by mimicking the appearance of host chicks. When little bronze-cuckoo (*Chalcites minutillus*) chicks hatch, they have the same dark black colouration as the hatchlings of their hosts, the large-billed gerygone (*Gerygone magnirostris*), whereas shining bronze-cuckoo hatchlings (*Chalcites lucidus*) are bright yellow, like the hatchlings of their yellow-rumped thornbill (*Acanthiza chrysorrhoa*) hosts. Horsfield's bronze cuckoos seem to mimic their hosts, superb fairy wrens (*Malurus cyaneus*), not just in appearance but also in the sounds that they make. These cuckoos flexibly adapt their call structure to resemble those being made by other nestlings in the brood. Australian scientist Naomi Langmore and her colleagues observed bronze cuckoos making short sharp chirps like the chicks of their fairy wren hosts. If they took one of these cuckoos and transferred it to the nest of another host species, the buff-rumped thornbill (*Acanthiza reguloides*), it only took a matter of days for the cuckoo to modify its call into drawn-out, lower frequency chirps like their thornbill hosts. These mimetic tricks are apparently convincing enough for the parents to not recognise that they are feeding a chick of a completely different species.

Mimicry of host chicks is sometimes seen in cases where the parasitic chick doesn't immediately or completely kick other eggs and nestlings out of the nest. In which case, blending in with the rest of the clutch seems to make sense. In other species, it's not that simple and the cuckoo chicks are reared alone or look conspicuously different to the host's chicks. The common cuckoo and their reed warbler hosts are a ridiculous example of this. Common cuckoos emerge from their eggs as small pink featherless hatchlings, but within a matter of weeks grow rapidly to dwarf the adult reed warbler in size. Fully grown reed warblers are about the size of a sparrow and weigh somewhere around 10–15 grams. By the time a cuckoo chick leaves the nest, it is almost fully grown and can weigh around 120 grams. The diminutive hosts continue feeding their enormous intrusive chicks even as they dwarf the reed warbler in size, and as they start to bulge out over the sides of the tiny reed warbler nest.

After the cuckoos leave their restrictively small nests, they remain dependent on their hosts for a few more weeks. By this stage, the chicks have developed the conspicuous banded feathers that make them look like a bird of prey and yet the reed warblers still diligently feed these strange nest invaders. The sight of a minute reed warbler feeding a screaming cuckoo chick ten times the size of its adopted parent is almost farcical and has fascinated scientists for generations. Many different ideas as to how this works have been discussed, including things like the cognitive abilities of some birds, the way that birds imprint on their offspring after they have hatched, and cost–benefit scenarios about whether a bird should reject or invest in chicks in its nest. In these cases, convincing a host bird to rear a cuckoo chick may be less about mimicry and more about tapping into and manipulating the parental instincts of the host.

As we've seen, the chirping cry of a hungry chick is an important signal that triggers the parent birds to behave in a particular way. Specifically, they gather food and feed it to the chirping chicks. This audio cue is usually accompanied by a bright conspicuous visual cue in the form of the bright yellow-orange open gape of the begging chick. This glowing beacon is an unambiguous bullseye for where an adult bird should put its gathered food. Cuckoo chicks also have brightly coloured gapes that they display when begging for food. This combination of nagging cries and bright yellow gaping mouth may be enough to tap into the parental instincts of the host and coerce them into feeding cuckoo chicks regardless of their overall appearance or similarity to nestlings of their own species.

This may only work for a short time as, like in the case of the common cuckoo, many cuckoo chicks are significantly larger than the chicks of their hosts. Run-of-the-mill begging behaviours may only result in the parent bird bringing enough food for a single chick of the same species. For this reason, cuckoos often exhibit exaggerated calls that coerce their hosts into bringing larger amounts of food as the chick grows bigger. This could mean faster, longer, more repetitive and more vigorous calling to convince their surrogate parents to forage more intensely. Big cuckoo chicks often have bigger and brighter open mouth gapes, which puts even more pressure on their hard-working hosts to collect more food. Scientists have wondered whether the rapid-fire calls of begging cuckoos worked by tricking the parent into thinking that they were hearing a whole clutch of nestlings begging for food, as opposed to just a single nestling, and thus foraging to feed an entire group. This doesn't seem to be the case, and it is more likely that the cuckoos' overblown calling behaviour taps into a more hardwired response, where exaggerated calling leads instinctively to extra feeding efforts from the host.

Horsfield's hawk cuckoo (*Hierococcyx hyperythrus*) elaborates on this and has evolved a startling way to convince their adopted parents to feed a single chick as if they were multiple chicks. Not only do they exaggerate their calls, but they have also evolved a way to increase the amount of visual signals they send their hosts. As the cuckoo chick chirps incessantly for food, it stretches out its wing exposing a bright yellow triangular patch on the underside. This triangle is the same colour as the bright yellow gape of the bird's open mouth. The Japanese scientists who discovered this suggested that this triangle fools the parent into thinking that it is an additional mouth waiting to be fed. Their evidence for this suggestion was pretty convincing, as the scientists observed the host bird, a red-flanked robin (*Tarsiger cyanurus*), trying to put food inside the yellow wing patch.

Cuckoos aren't the only birds that have evolved nest parasitism as their sole parenting strategy. Cowbirds (genus *Molothrus*), certain finches (family Viduidae), honey-guides (family Indicatoridae), and the black-headed duck (*Heteronetta atricapilla*) all must use other species of birds as surrogate parents for their offspring. Similar strategies have evolved in these birds as have evolved in cuckoos. Nestlings of screaming cowbirds (*Molothrus rufoaxillaris*), for example, have distinctively coloured feathers that make them remarkable mimics of the nestlings of their host species, the grayish baywing (*Agelaioides badius*). Cuckoo finches get their common name from being a finch that parasitises other species' nests. Adult female cuckoo finches (*Anomalospiza imberbis*) are remarkable mimics of another type of non-parasitic bird, *Euplectes* weavers. They are such convincing mimics that ornithologists erroneously classified them as a type of weaver, until a genetic analysis in 2001 revealed them to be another species of parasitic Viduidae finch. Resembling a harmless weaver, the cuckoo finch can avoid being mobbed defensively by their hosts, the tawny-flanked prinia (*Prinia subflava*).

While brood parasitism isn't restricted to cuckoos, they get the lion's share of the blame and reputation for this manipulative way of life. Also, let's not forget that around

60 per cent of cuckoo species do make an honest living by raising their own young. Despite this, the word 'cuckoo' has become synonymous with brood parasitism. So much so that it is added as a prefix to a long list of names of other animals known to be brood parasites. Cuckoo wasps, cuckoo bumblebees and cuckoo ants all have their own ways of convincing other species into carrying the burden of rearing their young. In the insect world, there are a few tricksters that are notorious for being just as deceptive as their avian counterparts.

Brood parasitism beyond birds

If you haven't noticed yet, butterflies and moths have been coming up a lot in this book. For almost every form of deception, there seems to be a moth or butterfly that has mastered it in some way. From the immaculate masquerade of the *Kallima* leaf butterflies, to the multiple mimics of toxic Amazonian butterflies, to sonar-deflecting luna moths, butterflies are among nature's most skilled tricksters. One potential reason for this may come down to the fact that moths and butterflies are, in general, frail, harmless, a little clumsy and very tasty snacks for other animals. Whereas some animals rely on a swathe of defences like strength, speed, smarts, venom or teeth, butterflies haven't evolved these strategies. There's the odd acrid-tasting butterfly here and there, but more often butterflies and moths rely on trickery for survival. Yes, they are lovely benign pollinators that deserve their status as beautiful wandering minstrels, but they're also lying bastards.

The alcon blue butterfly (*Phenagris* sp.) lays its eggs on the tips of marsh plants. The caterpillars hatch out and start feeding on the plant, just as you would expect any normal caterpillar to do. After a short time, the caterpillars let go of the plant and fall to the ground. For many other caterpillars, this would be a dangerous situation. On the ground, they can't access food and are vulnerable to being eaten by ground-dwelling predators like voracious ants. But for the alcon blue, this isn't an issue. When a worker ant from a *Myrmica* colony comes across a stranded alcon blue caterpillar, they pick it up in their jaws and carry it into their nest. Instead of carrying the caterpillar into the part of the nest where they store food (as they would be expected to do with any other stranded grub they find), they bring the alcon blue caterpillar into the nursery. Alcon blue caterpillars synthesise a chemical that makes them smell like ant larvae. The worker ants, upon finding a seemingly lost larva, diligently bring it back home where it is cared for and fed by the worker ants, along with the rest of the ant larvae. The caterpillars can stay protected and cared for in the ants' nest for up to two years until they are ready to pupate.

Just like in some cuckoos, the baby caterpillars don't look anything like their hosts' offspring, and they can rapidly grow to dwarf their adopted siblings in size. Again, just like in cuckoos, the caterpillars have ways to convince their surrogate parents to prioritise feeding them over feeding real ant larvae. To do this, the alcon blue shifts is deceptive strategy from mimicking an ant larva, to mimicking a queen ant.

Queen ants communicate with their workers using chemicals and sound. In *Myrmica* ants, the queens make distinctive pulsing vibrations that communicate their high social status to the worker ants. Alcon blue caterpillars make similar sounds which place them higher in the social pecking order of the nest than the other larvae (that don't make any sound), and this results in them being preferentially given food and offered greater protection by worker ants.

Hacking the chemical code of an ant colony to pull a deceptive trick like this is no mean feat. Ants use subtle chemical cues to distinguish nestmates from intruders. This includes recognising individuals of another species, but also ants from the same species intruding from neighbouring colonies. To overcome these defences requires that tricksters like the alcon blue use a combination of synthesising their own versions of ant-like chemicals and absorbing chemicals from the nest environment in a blend of mimicry and chemical camouflage. For something like a butterfly, which is distantly related to ants, to convincingly mimic ant signals is an impressive ability.

Other brood parasites play similar tricks to access the nesting areas of their hosts' colonies. The cuckoo wasp (*Hedychrum rutilans*) synthesises chemicals that mimic the scent of another wasp host species, the beewolf (*Philanthus triangulum*). With this chemical disguise, the cuckoo wasp can infiltrate the nests of their usually aggressive host species and lay their own eggs. Once the eggs hatch, the cuckoo wasp larvae start feeding on nearby beewolf larvae. There are ants that mimic the chemical signatures of other ant species, bees that mimic other bee species, and pseudoscorpions that dupe other pseudoscorpions, all so that they can wander into another species' colony where they can lay their own parasitic eggs.

Perhaps the strangest form of brood parasitism in the insect world is the case of a fungus that uses mimicry to infiltrate the broods of a termite colony. Scientists in Japan found small balls of fungus inside the nests of *Reticulitermes* termite nests clustered together with the termites' eggs. Normally termites would maintain a clean nest and clear any potential pathogens, including fungi, from their eggs. These fungus balls were small, hard, spherical, had unusually smooth surfaces and were mixed in with termite eggs in large numbers. This led the scientists to question whether these small round balls of fungus were showing some kind of egg mimicry. Like the animal nest invaders above, the fungus also appears to have a parasitic relationship with its host. The termites spend large amounts of time and energy caring for the fungus balls just as they would care for their eggs, keeping them clean and protecting them inside the nest from any would-be predators. What happens to the fungus from here on in is a mystery, and we don't know much about its life cycle, but there seem to be no benefits at all to the termites for keeping these egg-like fungus balls inside their nests.

Brood parasites present fascinating examples of deception in nature that are sometimes hard to believe. Their way of life relies on having evolved a specific relationship with another host species. They use deception to hack into and disrupt the life cycle of another species. Because brood parasites and their hosts have intertwining evolutionary histories they are engaged in evolutionary arms races, where the parasites are under pressure to evolve more sophisticated and convincing adaptations for exploiting their hosts. For each of the deceptive tricks of the various brood parasites we have talked about here, their varied hosts may also have their own bag of tricks to overcome and outwit parasitic nest intruders.

Similar dynamics are seen in other kinds of parasitic deception, where one organism hacks into the life cycle of another. Blister beetles present some of the most bizarre examples of this. Blister beetles are well known in some parts of the world for their bright aposematic colours. They produce toxic chemicals that can cause blisters on people's skin, hence their common name. However, in some cases, the animals most vulnerable to blister beetles are solitary bees. In these species, blister beetle grubs invade the nests of solitary bees and feed on the bee's eggs and pollen stores. They don't sneak inside hives like alcon blue butterflies, they are delivered directly into the nest by the mother bee. The beetles manage to do this thanks to a deceptive life cycle that requires they become some strange form of collectively mimetic sexually transmitted infection.

Meloe franciscanus blister beetles lay their eggs in the ground in large clutches. When the eggs hatch, the grubs emerge and, as a group, begin climbing up nearby vegetation. They aggregate together in a dense ball of tiny crawling grubs on the ends of twigs or long grass stalks. These aggregations can be made of anywhere from a few hundred to a few thousand individuals. Together they look like what you would expect – a swarming ball of reddish-brown wriggly things. But they smell like something completely different. As a group, the beetle larvae emit pheromones that mimic female white-faced bees (*Habropoda pallida*). This is a solitary bee species, where males and females can be found flying around flowers searching for food, or in the case of males searching for females to mate with. Female bees often perch on the tips of twigs and grass stems where males will fly in and attempt to mate by grasping onto the female from behind. Lured by the scent of an aggregation of blister beetle larvae, male bees also fly close and land on the aggregation. As soon as the bee has made contact, the beetle larvae crawl en masse onto the bee's underside. In an instant, the bee can be swarming with several hundred tiny beetle grubs. They hold on tight, clinging to the bee's hairy body as it flies away. Then, when the male finally finds a real female, and lands on her back, the beetle larvae quickly jump ship. They crawl from the underside of the male and latch onto the back of the female. Again, they hold on tight to the back of the female bee until she eventually returns to her nest. Here the blister beetle larvae disembark, and the carnage begins. Safe inside the bee's nest, which is buried deep underground, the beetles have a store of food that can last them several months until they finally pupate and emerge from the bee's nest as adult beetles.

Animals that use deception as a crucial part of their complex life cycles seem to use strategies that rely on so many different moving parts falling into place. Not only must the deceivers look, smell or sound a particular way, but they must also behave in a particular way. The smell of a cluster of blister beetles is wonderfully deceptive, but it must also be paired with the appropriate swarming behaviours for the blister beetle's deceptive life cycle to complete. Similarly, the deceptive calls and colours of cuckoos align with other behaviours, such as appropriate nest choice by the egg-laying adults or egg-eviction behaviours of the cuckoo hatchlings.

For most of this book, we have focused on the anatomy of creatures; their colours, shapes, smells and sounds, and how these can mislead and misdirect. Specific behaviours can sometimes go along with these deceptive cues, from the wriggling tongues of snapping turtles to the choice of an appropriate background for a camouflaged animal. Behaviour is a tricky subject when it comes to understanding deception because we often interpret behaviour as being a result of thought. The mind thinks something and then the body behaves in a particular way because of that thought. This starts to raise prickly questions about those thoughts. How conscious are these thoughts? How intentional are they? And if these thoughts involve deception, do animal minds have any conscious appreciation of their own deceptive nature? Or put in simpler terms, do animals know that they are tricking other animals? It's a difficult question, and one that scientists are only beginning to dare answer.

8

Can animals tell lies?

I've spent a lot of time giving public talks about deceptive animals and the tricks they play. Most of these talks focused on my research into orchid mantises and how they lure in pollinators because of their flower-like appearance. I'd display colourful photos of mantises against real flowers to show their similarities, tastefully designed presentation slides with all manner of bar charts and scatterplots, and slow-motion videos of bees zipping back and forth in front of orchid mantises as they flew headfirst towards their demise. I would get the polite smiles and nods you'd expect at a talk from a university boffin. Sometimes I would get audible 'oohs' and 'aahs,' as people watched footage of mantises grabbing insects out of mid-air and eating them alive. At the end of each talk, there would be time for questions. After having given so many of these talks, I got pretty good at predicting what the questions would be, and had practised and prepared answers ready.

Q. *Can they change colour?*
A. *Yes, they can!*

Q. *Do they smell like a flower?*

A. *No, they don't; there might be some scent cues involved, but it's still being figured out.*

Q. *Can you keep them as a pet?*

A. *Yes, many people do, but they are a protected species so it's probably not a great idea.*

You get the gist.

One day, I was giving a talk at a senior citizens club who had asked me to visit and tell them about the work I did as a scientist. I got the expected 'oohs' and 'aahs', even a few polite laughs at some of my cornier jokes and truly awful puns. It was all going well until the end, when I was asked a question that I had never been asked before and that left me completely stumped. Not because I didn't know the answer, but because the question was so unexpected I didn't even know where to begin answering. A lady put up her hand, and said, 'How does the mantis know to look like a flower?' I paused to think, and eventually said that I didn't understand the question and asked if she could repeat it for me. She went on to say that she was completely fascinated by the idea that this mantis could trick bees into flying towards it, but was baffled as to how the mantis *knew* what to do to trick bees. *How did it know to look like a flower?*

Completely unprepared for this question, I just said: 'Well … it doesn't.' I then anxiously and disjointedly muttered something about evolution, natural selection and blind watchmakers, before finishing timidly with, 'Does that make sense?' Her smiling silent response was something between a head-nod and a shoulder shrug, which I assume meant that it didn't make sense at all, but it didn't matter and could I please stop talking now. Having had some time to think about it since, it's clear that this question came from a different perspective than I was used to.

This person was not a biologist and may not have come to the talk with any prior interest in science or natural history. My casual stories about animal tricksters and their intriguing behaviours were then interpreted very differently from how I had assumed they would be. The phenomenon of deception, as far as this person may have been concerned, was something intentional. In humans, deception can take many forms: lies, fraud, con games, rigged systems and stacked decks. At its worst, deception implies people knowingly misleading others. As I was telling a story of orchid mantises deceiving pollinators, this person was hearing a story about intention, where conniving mantises were consciously pulling-one-over on innocent bees.

Throughout this book, I've playfully described the tricksters and charlatans of the natural world who survive by tricking other animals into behaving in particular ways. As fun as it can be, I understand all these terms come with rather negative implications. To deceive, dupe, trick or outwit sounds nefarious. This is partly a limitation of language. I challenge you to go through your thesauruses and find synonyms that don't give off similarly negative vibes.

To be clear, most of the trickster animals and plants described in this book have no idea that any deception is taking place.* There is no duplicitous intent behind their

---

* And I must apologise for calling butterflies 'lying bastards' in the previous chapter. I may have gone a little too far with that one.

interactions with other animals and it's very unlikely that the deceptive creatures perceive their own appearance or behaviours in the same way that their victims do. An orchid mantis doesn't need to know what a flower is for their pollinator-luring strategy to work. Like I described in chapter 5, you can make plasticine models of orchid mantises that are just as good at luring bees as real orchid mantises. The trick still works without there being any possible intention or knowledge on the part of the deceiver. Similarly, a hoverfly doesn't need to know what a bee is to be a convincing mimic of one. A rewardless orchid has no concept of honesty versus dishonesty and thankfully has no idea what those randy wasps are doing to its flowers. These are all adaptations for survival that have arisen through evolution via natural selection. The physical structure, scents and sounds of those organisms have been finely tuned through generations to look, smell and sound like other things. Not through any conscious effort but simply because their similarity gives that organism an upper hand in the games of survival and reproduction. Repeat this for generations, and evolution gives us some mind-boggling cases of mimicry and other bizarre similarities.

This is true, for the most part, throughout this book. Camouflage, mimicry and deceptive pollination all work without any conscious trickery going on. However, in the last two chapters we started encountering some more questionable animal behaviours where this might not necessarily be the case. Mourning cuttlefish quickly change the colour and pattern of their mating display depending on where other males are in the vicinity. Nursery web spiders conceal worthless gifts in layers of silk to give to females. These kinds of tricks aren't necessarily an inherent part of the way that animal's body is built. They have an air of mischievousness about them, as if they were making a choice to deceive another animal. These sorts of behaviours start to raise a very tricky often controversial question: can animals tell lies?

The truth about lies

To knowingly mislead another is often characterised as a uniquely human ability. Scientists, psychologists, sociologists, economists and more have all turned their attention towards trying to figure out why and how we conceal the truth from others for our own benefit. Perhaps unsurprisingly, revealing the truth about our tendency to conceal truths is quite difficult. It's generally accepted that lying requires the brain to do some rather complex processing. It requires that people have the ability to imagine what another person is thinking, including the possibility that they may be thinking something different to themselves. Compared to telling the truth, extra brain power would be needed to, firstly, be aware of a truth; secondly, invent a plausible re-creation of the truth; and then convincingly communicate the imagined version to another person. Add to this the consideration of social consequences for transgression, and the simple act of telling a lie starts to sound like high-level mental gymnastics. Brain scan studies seem to confirm this, and people appear to exhibit much higher levels of brain activity when crafting lies than when telling the truth.

Attempts at developing techniques to detect lies are generally problematic. Polygraph lie detector machines, which you've no doubt seen in old thriller movies, are the modern equivalent of a centuries-old tradition of inventing convoluted and controversial methods to glimpse inside a person's mind. Ancient Greeks attempted to detect lies by measuring a person's pulse. Stories from ancient China tell of accused liars having their mouths filled with dry rice. If the rice they spat back out was still dry, then they must be a liar. These are the kinds of physiological reactions we expect from a caricature of a fearful liar wracked with guilt – a dry-mouthed person pulling nervously at their collar and mopping their sweaty brow.

Modern-day polygraphs refine this approach and presume that the stress and cognitive load of crafting lies should be accompanied by subconscious physiological responses like increased heart rate, respiration, perspiration and blood pressure. By placing sensors on a person's body that measure these physiological responses while being asked a series of questions, it's expected that these subconscious reflexes will reveal the truth or untruths hidden behind the answers that the person gives. However, the list of confounding factors that could interfere with physical symptoms is so long it easily casts a shadow of doubt over every test result. Everything from alcohol to prescription medication to arthritis can affect our nervous systems enough to interfere with polygraph readings. The simple conundrum that a person's fear of a polygraph machine, whether they are guilty or innocent, could interfere with their physiological responses, makes the interpretation of lie detector results immediately questionable. Numerous studies have drawn attention to remarkably high error rates in the accuracy of polygraph tests and there is a general consensus among psychologists that they are of little use beyond being large and expensive doorstops.

Humans are even worse lie detectors. We might pride ourselves on our ability to detect subtle behaviours that reveal a person's true character, but empirical research shows that we may as well be flipping coins to decide whether someone is telling the truth. In 2006, two researchers, Charles Bond and Bella de Paulo, compiled decades of research into how accurate people were at detecting when another person was either lying or telling the truth. After reviewing data from tens of thousands of tests, they concluded that people got the answer right about 54 per cent of the time. Like I said, we may as well be flipping a coin. Supposed 'expert' lie detectors, like police officers, judges and psychiatrists, had a success rate about a percentage point lower than non-experts. Add to this list of problems issues around defining what counts as a lie and what doesn't. White lies (like the ones about your kid's violin playing or your aunt's new tuna bake recipe) are discounted as even being deceptive in the first place and accepted as normal social graces.

When we can't detect lying in the person sitting across the table from us, how could we ever hope to understand whether it's happening in a cuttlefish? Whenever scientists go looking for human-like traits in other animals, they inevitably start by looking at apes and monkeys. As our closest animal relatives, we share many characteristics. If conscious deception happens in our nearest relatives, it might shed some light on whether lying is uniquely human or a trait we inherited from our ancestors long ago.

In a classic study from 1979, scientists tested whether chimpanzees (*Pan troglodytes*), our closest living relatives, were capable of intentionally misleading humans. Captive chimpanzees were placed in a small enclosure. A person would then place two boxes outside the enclosure, one with food in it and one without. A different person would enter the room and for the chimpanzee to get the food, they would need to gesture towards the box containing the food. If they pointed to the correct box, the person would take the food out and give it to the chimp. The chimpanzees quickly got the hang of this task and, over repeated trials, would readily point towards the correct box.

This was straightforward until the scientists introduced another person into the experiment – one that was competitive instead of cooperative. Instead of handing the food over to the chimpanzee, this person would take the food out and keep it to themselves. The trials would alternate between cooperative and competitive handlers, and the scientists made it easy for the chimps to distinguish between the two. In a rather pantomime fashion, the competitive handlers were dressed in black boots, dark sunglasses, and a bandit mask. Eventually, after enough experience, the chimps started behaving differently towards the shady-looking food-stealers. Instead of pointing to the correct box, they started gesturing towards the empty box. Instead of simply ignoring the competitive person, or refusing to participate, the chimps successfully completed the task of gesturing towards one of the boxes but would give misleading information to some people and not others.

This type of behaviour has been termed 'tactical deception' to distinguish it from other forms of deception. It implies that the animal understands that it can transmit information to other animals and chose to transmit incorrect information to mislead others. This study set a benchmark for understanding whether animals could intentionally mislead and was, at the time, a great challenge to the idea that humans have a special kind of intellect that allows us to tell lies. Since then, similar studies of captive animals have shown that other primates – such as capuchins, squirrel monkeys and lemurs – display at least some level of similarly deceptive behaviour. Scientists studying free-ranging primates, such as mangabeys (*Cercocebus torquatus torquatus*) and Tonkean macaques (*Macaca tonkeana*), have witnessed similar behaviours being used between primates within their own social groups, showing that tactical deception isn't confined to captive animals interacting with humans.

Those who believed that humans were exceptional in our cleverness and ability to conjure tricks are having to eat their words. Though there is still a prevailing view that tactical deception must be the result of advanced intelligence and that we, the cleverest of all apes, must therefore be the best at telling lies. The neocortex is the part of the brain that handles higher-order brain functions like problem-solving, decision-making and processing sensory information. It's the wibbly-wobbly bit on the outside of our brains and it is much larger in humans and other primates compared to other mammals. Our large and complex neocortices are understood to be the reason that primates are so smart and can do clever things like use tools, do basic arithmetic, wear cargo pants and play minigolf. One study of primate brains showed that relative neocortex size has a strong correlation with how often different species use deception in social contexts. This further supports the idea that tactical deception takes smarts and perhaps makes us feel a bit better about our own tendencies to lie, cheat and steal. We're not bad people; we're just burdened by our immense intellect.

But before we get too cocky about our bulging neo-cortices and prodigious capacity for crafting untruths, primates aren't the only animals that can spin a tall tale. In 2017, a team of scientists repeated the chimpanzee study above with another clever type of animal: domestic dogs. In a modified version of the same experiment, dogs were trained to lead their handlers towards boxes that either did or didn't have food in them. In this study, they didn't bother dressing people up in bandit masks and sunglasses; the only consistent difference between cooperative handlers and competitive ones were that the cooperative people gave the food inside the boxes to the dog, and the competitive people kept the food for themselves. After a few training sessions, the dogs were tested to see how they treated these different people. Just like in the chimpanzee study, the dogs were more likely to lead cooperative people to the box with food, and competitive people to the empty box. Not only were dogs capable of misleading certain people over others, but they learned to do it incredibly fast. In this study, the dogs were trained six times with a combination of cooperative and competitive handlers. After this small number of trials, the dogs were able to distinguish the two people and effectively mislead the competitive handlers. In the chimpanzee study it took months to years and sometimes several hundred experimental trials before the chimps could reliably mislead their competitive counterparts. It turns out these big-brained chimps might not have been so clever after all.

A big neocortex isn't just a tool for solving basic arithmetic. One of the best predictors of neocortex size is the size of an animal's social group. Bigger social groups mean bigger neocortices. That tactical deception has evolved in monkeys and apes may have less to do with their problem-solving skills, and more to do with the fact that they are social animals that live in tight-knit groups. The cognitive and social skills required for animals to live in groups may also provide the perfect toolkit for tactical deception. This may go some way to explaining why liars and cheats prosper in many different types of social animals, including primates, dogs, birds, even rodents.

Who's watching your nuts?

People reading this in the northern hemisphere may be very familiar with squirrels. Perhaps a little too familiar with squirrels. You may have seen squirrels so often, and become so accustomed to their presence, that you have become desensitised to their charm and brilliance. For many people, squirrels are just sewer rats with fluffy tails, garden pests and invaders of roof awnings. However, for an antipodean like me, squirrels are majestic creatures from another universe. I watch in captivation as they pounce weightlessly, chased by their impossibly delicate tails twisting and curling like amber mist. I've seen more cartoon drawings of squirrels in waistcoats and peaked caps than I have seen real live squirrels. So, on the few occasions when I've seen squirrels in the wild, I've become intoxicated by their adorable little faces and tiny little hands that have somehow jumped out of myth and fable and into the flesh before my eyes. Yes, I have been that tourist fixedly crawling on all fours with an oversized camera lens, trying to get the perfect wildlife shot of what to many others is nothing but a common urban nuisance.

The same phenomenon happens in reverse when people come to visit Australia, and I can't quite share their enthusiasm for seeing yet another kangaroo plodding about munching grass. Don't get me wrong, I think kangaroos are great, but I've seen enough in my time that I would somehow need to see them with fresh eyes to share others' enthusiasm for how bizarre and beautiful they truly are. Squirrels, on the other hand – gee-whiz, those adorable little nuggets just need to twitch their tiny whiskers and my heart melts. And don't get me started on those cheeks! How can you not giggle with delight, seeing them puff up their teeny faces with goodness-knows-how-many nuts and seeds they can fit in there. The first time I travelled to America, I must have spent an entire day sitting in the middle of a field in Central Park, New York, just watching little grey squirrels dart about, filling their pudgy cheeks with food scraps and disappearing up nearby tree trunks.

Squirrels don't just hide food in their cheeks, they hide it everywhere. Rather than just eating food as they find it, squirrels will take things like nuts and wedge them inside tree hollows or bury them underground for later. Instead of hiding all their food in the one spot, they hedge their bets and scatter their food all over the place. They have an amazing spatial memory and can remember hundreds of individual food caches spaced out across several hectares. This helps them survive the long winter months when they can revisit their food caches and feed on things like nuts stored safely inside their hard shells.

One drawback to having food stores scattered all over the place is that food can be stolen by other animals while the squirrel is away. Other squirrels and clever birds like crows are known to quietly watch as a squirrel is hoarding their food. Then, once the squirrel runs off, the onlookers can sneak in to steal the stored food. While stealing another animal's food is sneaky, it's not necessarily deceptive. If you steal someone else's food while their back is turned, it doesn't require you to manipulate their behaviour in any way. It requires you to be a bit of a jerk, but not a dishonest jerk. No, the trick here is played by the squirrel that is hiding its food.

When squirrels store food, they start by collecting something like an acorn and holding it in their cheek pouches. They go looking for a good patch of earth to store their food, then dig a small hole. Once the hole is deep enough, they reach in with their heads to push the acorn down and then cover the hole back up again with dirt. But occasionally they do something strange – they dig a hole, stick their heads in, and then cover the hole up again without actually burying the acorn. Perhaps they changed their minds and decided that this hole wasn't a great spot for their acorns. This doesn't explain why they would go to the effort of filling the hole back up again. Do they just appreciate a tidy garden?

Scientists observing this behaviour started to wonder whether something else was going on, and whether squirrels were deliberately pretending to bury nuts to mislead other animals. The team of scientists studied eastern grey squirrels (*Sciurus carolinensis*) living freely in parklands in North America. For their experiment, they would pick one focal squirrel, then observe their behaviour and record whether they cached their food or 'pretended' to cache their food. They also counted how many other squirrels there were in the vicinity and how close they were to the focal squirrel. The pattern was pretty clear: when more squirrels were around, and when other squirrels were closer to the focal squirrel, the focal squirrel was more likely to show fake food-hoarding behaviours. It looked like the squirrels were paying attention to where and how many other squirrels were in their environment, and changing their behaviour to suit.

The scientists messed with the squirrels' heads a little bit more by going around and stealing food from their hiding places. Whenever squirrels experienced having their food stolen, they started doing more of the same deceptive burying behaviour, apparently in an attempt to mislead nearby animals as to where their food was hidden. And it seemed to work too. When scientists went looking for hidden squirrel food, they had a much harder time finding it when the squirrels made fake caches, compared to when squirrels didn't show these deceptive behaviours. Their pretend nut-burying behaviour seems convincing enough to fool humans, as well as other nearby squirrels.

Family ties lead to family lies

Survival in social groups depends on communication and cooperation. When groups function well, individuals can share the load of responsibilities, like gathering food and defending themselves from predators. A classic example is meerkats (*Suricata suricatta*). If there were ever an animal that rivalled squirrels in their undeniable cuteness, it might just be meerkats. Admit it, you can't help but be enamoured of the way they stand up on their hindlegs and look about with their big glossy black eyes. If you've ever seen them at the zoo, you've probably spent a silly amount of time watching that one adorable meerkat slowly nodding off to sleep when it should be looking about for predators. Meerkats mostly feed on insects and grubs that they find buried underground, which makes keeping an eye out for predators difficult – it's hard to spot an eagle flying above when your head is probing around under the soil. Group living solves this problem and is why group members take turns in the role of sentry: staying on the surface and watching for predators while others focus on hunting for food. If one of the sentinel meerkats spots a predator, they let the group know with a loud alarm call. They have specific types of alarm calls for aerial predators and ground-based predators. When the call is heard, the foraging meerkats abandon whatever they are doing and flee to safety with the rest of the group. Whether in birds, rodents, primates or any other type of social animal, similar communication channels exist to share information between individuals of a group.

For these communication systems to work requires a high degree of trust between individuals in a group. It's easy to imagine, though, how such systems would be vulnerable to exploitation. Aesop's classic fable 'The Boy Who Cried Wolf' is a cautionary tale about the dangers of dishonesty. In this fable, a young shepherd is placed in charge of a flock of sheep. When he cries, 'Help, there's a wolf!', the villagers come running over to help, only to discover that there is no wolf to be found. In some versions of the story, the boy is motivated to do this by fear, in others by boredom, and in others he simply gets a kick out of manipulating people with such a simple trick. This same kind of trickery happens in social animals, but with one key difference – in the fable, the boy cries wolf to draw other group members near; in animals, they cry wolf to make group members scatter and flee.

Golden snub-nosed monkeys (*Rhinopithecus roxellana*) live in small groups of usually fewer than ten individuals. They spend most of their time in trees but will come down to the forest floor as a group to feed. They're more vulnerable to predators while on the ground, so the group must keep watch. If one monkey spots a predator and makes an alarm call, known as a 'chuck' call, the troop must scamper back to the trees, leaving their food behind. A team of scientists in China set up cameras to remotely observe the behaviour of wild golden snub-nosed monkeys in the Guanyinshan Nature Reserve. When they examined how the monkeys used their alarm calls, they found that over half of all alarm calls were false alarms. A monkey would make the 'chuck' call, and the group would scamper to safety despite there being no predator to be seen. Meanwhile, the monkey that made the alarm call would stay on the ground and run around picking up the food left behind by the other monkeys. They were making false alarm calls to scare off the monkeys in their own social group and steal their food. By observing golden snub-nosed monkeys over long time periods, the scientists discovered that the monkeys seemed to use this trick more often when food was scarce and competition was high, such as during the cold winter months. Similar behaviour has been observed in other animals that seem to use the same trick to steal food from other members in their group. It's been put forward as an explanation for false alarm calls in everything from tufted capuchins, to great tits, to Arctic foxes.

The fable of the boy who cried wolf isn't just about how people can be manipulated with lies, it's about the disastrous repercussions of lying. In the fable, the boy cries wolf a second time, and the villagers rush over to the boy and his sheep to discover once again that there is no wolf. Soon after, a real wolf appears. Once again, the boy cries 'Help, a wolf!' But this time, the disgruntled villagers stay home, and the wolf eats the entire flock of sheep. Just like in the fable, there are likely to be consequences for animals who lie too often. In snub-nosed monkeys, the more recently a group member has made a false alarm call, the less likely it is that group members will respond to another alarm call. This is potentially disastrous for the group if the second alarm call is a real one. At best, it's a downer for the monkey attempting to get an easy meal with another false alarm call, only to discover the group is onto their ruse. One animal has figured out a way to overcome this problem and prolong the game of 'cry wolf' before getting ousted as a liar.

Fork-tailed drongos (*Dicrurus adsimilis*) are small black birds that live in the arid African savannahs and are becoming famous for being some of the most cunning liars in the animal kingdom. They don't fool other drongos; they have ways to trick other family groups from a completely different species. Their unfortunate victims are none other than those poor, sweet, innocent, glossy-eyed meerkats. When meerkats listen out for alarm calls, they don't just listen out for calls from their own species, they listen out for alarm calls from nearby birds as well. Fork-tailed drongos are often found hanging around near meerkat families. They have a range of distinctive raspy 'chink' alarm calls that they make when they spot a predator. Meerkats recognise the drongos' call and will drop what they are doing to run for cover. This is a useful way for meerkats to increase the number of watchful eyes keeping their family group safe. But it also makes them vulnerable to crafty drongos who make false alarm calls that send the meerkats scampering. Once the coast is clear, the drongo will fly in and scoop up any food that the meerkats have abandoned. This way, the drongo benefits from all the hard work the meerkats have done by digging up worms and insects from underground.

The meerkats quickly catch on though. If a drongo tries the same trick again, the meerkats are less likely to respond. So, the drongo mixes things up a bit and tries a new type of alarm call – a meerkat alarm call. Fork-tailed drongos are remarkable vocal mimics and can do an uncanny impression of a meerkat alarm call. The meerkats respond to it just as they would one of their own and scamper to safety. Meanwhile the drongo flies in once more to steal the meerkats' abandoned food. They don't stop there; drongos mimic a whole range of other animals' alarm calls, including other nearby birds such as pied babblers and caped glossy starlings. The hypervigilant meerkats are again tricked by these calls into abandoning their food. One study counted 51 different types of alarm calls made by fork-tailed drongos. By switching up their cry-wolf repertoire, fork-tailed drongos can avoid meerkats becoming desensitised to their alarm calls, and get away with stealing more food. And if all else fails, fork-tailed drongos don't just rely on meerkats to do the hard work. Their vocal mimicry tricks work on other animals, too. Sociable weavers and pied babblers are both group-living birds that forage in flocks on the ground. Drongos can use false alarm calls to trick these birds into abandoning their food as well. If only the boy who cried wolf had the ingenuity to switch tactics every now and again, he could have gotten away with his tomfoolery. Though it would have made for a very different fable and I'm not sure what the moral of that story would be.

Yes or no, can animals tell lies?

It's early days in our understanding of whether animals can tell lies. Social group-living animals are likely candidates for finding animal tricksters in the wild, and we have only scratched the surface of what kinds of trickery goes on in these societies. Time will tell whether similar forms of deception are happening in the subsonic songs of whales and dolphins, the chemical mists of ants and termites, or the barks and howls of wild dogs. Despite the growing list of evidence, scientists are still hesitant to conclude definitively that animals can intentionally deceive another, in a similar way to humans. This might have more to do with the professional conduct of scientists than it has to do with the reality of what goes on in animal societies. Firstly, one of the most important rules in studying animal behaviour is to not anthropomorphise animals. In other words, don't project human-like qualities or emotions onto other species. Scientists must be completely objective and not assume that other animals think, behave or make decisions for the same reasons we do. Since lying has always been perceived as a human behaviour, scientists are incredibly hesitant to conclude that it could be happening in other animals.

The second reason is that scientists are trained to never definitively conclude anything. There is an idea in science that you can *disprove* things, but never really *prove* things. This is why you will find scientific writing littered with phrases like 'the evidence suggests', 'it's probable that', or – my personal favourite – 'the most parsimonious explanation is', and so on and so forth. A professional scientist will often give a direct 'No' answer but rarely a direct 'Yes'. It's like squeezing blood from a stone. But this is a good thing. What often sounds like indecisiveness or a lack of certainty is actually just good professional practice. It's part of a scientist's job description – to consider all possible explanations for the observations they have made and examine every possibility on its own merits. Then the conclusions that are drawn come down to which of these possibilities is *most likely*. In a universe of infinite possibilities and scales of enquiry, it would take an impossible amount of overwhelming data for scientists to exclude all possibilities but one.

Let's use false alarm calls in golden snub-nosed monkeys as an example. One explanation may be that they are knowingly misleading their group members to get food, another is that false alarm calls are a learned behaviour that the monkeys use to get food without an understanding that it's being used dishonestly. What would each of these scenarios look like in reality?

For the first scenario, imagine a monkey sitting on the ground in a forest foraging for food in leaf litter with its troop. It understands the meaning behind a predator alarm call and anticipates that if it makes an alarm call, the other members of their troop will scamper up trees leaving their food behind. With this knowledge in mind, it makes a deliberate decision to use a predator alarm call as a tactic to get food. The monkey understands the mindset of the other monkeys nearby and intentionally makes the alarm call in the incorrect context, predicting that it will have the effect of scaring the other monkeys away. This would count as tactical deception and if we knew for certain this was the monkey's thought process, we could conclude as such.

The second scenario posits a situation where a monkey learns that a particular call type results in it getting food, without understanding that the other monkeys perceive that alarm call as having a different meaning. Picture that monkey back on the ground foraging with its troop. Imagine that, for some reason, it accidentally makes an alarm call or something that sounds like an alarm call. Maybe it mistakenly thought it saw a predator or just happened to vocalise in a way that was somewhat like an alarm call. The next thing the monkey sees is the rest of its troop scurry away, leaving perfectly good food behind. What luck! Repeat this mistake a couple of times and the monkey soon learns that if it makes a particular sound, it can get the food that the nearby monkeys have. It develops an understanding that there is this one call that can be used in two different scenarios. One, when you see a predator, and two, when you want the food the monkey next to you has. So, the monkey starts making false alarm calls and then goes scurrying around, picking up the leftover food and wondering why Susan and Margaret are busy freaking out halfway up a tree trunk.

In both scenarios, the behaviour looks the same. What we can't see is what is happening inside the minds of the animals. We could create similar thought experiments about drongo alarm calls or scatter-hoarding squirrels. What do you think is happening here? What explanation do you think is *most likely*? We may never be able to design an experiment that clearly teases the two possibilities apart. So, we are left to decide based on incomplete data between the two possibilities.

The scientific practice of being open to all possibilities is an admirable quality. To be open to being proven wrong, to question tightly held beliefs and to always be prepared for greater planes of awareness is what drives science forward. They are also (I believe) tendencies that we could all benefit from in our daily lives. However, scientists are only human, and even though professional practice prohibits them from concluding definitively that animals can tell lies, behind the scenes, in unspoken thoughts, they have a good idea of what's going on. While you might never find a straight answer in a scientific paper, find those scientists off the clock, maybe buy them a couple of drinks, and ask them the question: 'Can animals tell lies?', you might (possibly, most likely, just maybe, in all likelihood) get a straight answer like, 'Yes, they can'.

9

Finding inspiration
in illusion and deception

We have all, at some point in our lives, been unlucky lovers, impulsive reactors or victims of our own limited eyesight, hearing and slow thinking. We know what it is to be deceived. While this book focuses on the inner worlds of animals and plants, hopefully at least some of these stories feel somewhat relatable to us humans. Underneath our polyester fabrics and silver jewellery we are, after all, animals. We're products of the same shared evolutionary history, and our minds and senses are as prone to the same forms of trickery as is any other animal. The same deceptive phenomena experienced by the plants and animals in this book can be seen in our daily lives. Not just in the tricks that we play on others or those that are thrust upon us. The same tricks used in nature are also used in art, be it painting, film, fashion or music. We can take these quirks of our biology and use them to tell stories and create new experiences.

May I, just briefly, go on one last strange tangent? Do you know what a three-dimensional zoetrope is? You might be more familiar with two-dimensional ones – these are ingeniously simple gadgets that illustrate how frame-by-frame animation works. Imagine a wide shallow cylinder with vertical narrow rectangles cut into the sides at regular intervals. On the inside surface of the cylinder, nestled in between the slender windows, are still frames from either a film or animation. Looking through the windows, you can see the still frames on the opposite side of the cylinder. The magic happens when you spin the cylinder. As one narrow opening in the cylinder zips past your eyes, you catch a quick glimpse of one still frame. When the next opening passes your line of sight, it reveals the next frame in sequence. The flicker of passing windows reveals frame after frame and, just like a reel of film, the images come to life and move. This illusion tricks our eyes in the same way that scientists have pondered when trying to make sense of banding patterns in fast-moving animals.

Three-dimensional zoetropes work on the same principle, except instead of using two-dimensional images, they use sculptures positioned in a circle on a spinning platform. Each sculpture in sequence is slightly different in pose or position to the one before it, just like the still frames. As the platform spins, a strobe light from above pulses. The strobe light controls the frame-rate at which you can view the zoetrope. If the timing is right, you will see a sculpture in a particular position during one flash of the strobe light, then during the next flash you will see the next sculpture in sequence in the same spot. This continues on in a loop; flash after flash, sculpture after sculpture. The resulting illusion is wonderful and a little disconcerting. Right before your eyes, inanimate objects become animate. It's the same principle behind stop-motion animation which, incidentally, is one of my favourite artforms. I love how inert blocks of clay, wood and wire can, with a few tricks of the eye, be brought to life to tell incredible stories.

I had no idea three-dimensional zoetropes existed until I stumbled across one at the Museum of Old and New Art (Mona) in Hobart, Tasmania. For a brief moment, I found myself gazing at a wonder that I couldn't quite explain. It gave the fine tether that connected me to reality a sudden unnerving shudder. The piece is called *Artifact*, created by artist Gregory Barsamian. I turned a corner in the dark galleries of the museum and was surprised to see an enormous bronze head lying on its side. Small windows cut into the head emitted an eerie flickering blue light. I walked up to the head and peered into one of the windows. Inside the glowing blue head were clay birds flying through the air before disappearing into the pages of floating books. Falling apples landing on dismembered hands before melting away and oozing between the fingers. It was clear that this wasn't a projection; the objects were there right in front of me and impossibly moving. I had no idea what I was watching. I didn't understand how inanimate objects were morphing and moving right in front of me. I felt a rush of adrenalin as I was confronted with something that didn't mesh with my understanding of reality. I was, in that instant, like James Hingston watching an orchid sprout legs and walk.

This moment of elated confusion may have only lasted a second or two, but it was jarring enough to be memorable. Eventually I started noticing some finer details. There was a tangle of black wires weaving between the objects that I realised were support structures keeping the objects suspended in the air. The strobe light, I figured, probably wasn't just an artistic choice; it was there to play tricks with my eyes. I put the pieces together and the method behind the illusion was revealed to me. This new level of understanding didn't lessen the magic of the artwork; it was enhanced. I laughed when I realised that I was witnessing stop-motion animation being played out in real time, right within my grasp.

Perhaps it was the ambience of a serene gallery, or the unsteadying flicker of the strobe light, that helped tip my mind into the realms of impossibility. Whatever happened, I loved that little moment of confusion when I couldn't make sense of the world right in front of my eyes. The unreal suddenly became real. I think this is why we are fascinated with deception. It's the same feeling that we get when we view a great work of art, are fooled by a simple magic trick, or experience a new piece of technology we can't yet comprehend. Realising that evolution has formed natural illusions all around us can give that same elating buzz of adrenalin. Finding the camouflaged animal hidden right in front of you, being fooled by the skilled mimicry of a deceptive bird call, or looking at a twig only to find that the twig is looking back at you ever so slightly shudders our footing with reality in a way that inspires intrigue and excitement.

These experiences, I feel, remind us that we are part of something greater. That we are not the centre of our own universes. We are organisms navigating our way through a world of molecules and photons, using a small assemblage of limited sense organs. Our eyes are imperfect windows into the world. Our minds are entirely satisfactory processors, but unreliable in their own right – not built through strategy, but through the dispassionate filter of our evolutionary past. Understanding this ties us back to a vividly real world full of orchids, butterflies, dark abyssal oceans and dense shadowy jungles. We are part of it, too. Wonderful creatures. Bodily humans.

The greatest shift in humanity's understanding of this came with the formation of evolutionary theory. When it was finally clear that the living things around us were all products of natural selection, the colours, shapes, sounds and smells of living things could be interpreted as having a role to play, and that role could be to deceive. From the mottled hues of a camouflaged coat, to the uncanny cry of a deceptive bird call, evolution could give us natural cases of dishonesty and deceit. Since the pioneering work of explorers and scientists like Charles Darwin, Alfred Russel Wallace and Henry Walter Bates, scientists have been finding more and more ways in which nature is never quite what it seems. The science of deception grew from plausible ideas to testable hypotheses thanks to the meticulous observations of people like Abbott Handerson Thayer and Hugh Bamford Cott. They built upon the evolutionary hypotheses of *why* camouflage and mimicry are happening, to formulate specific hypotheses of *how* it happens. Their ideas around how colours, tones and shapes can create illusions in the minds of other animals are still being tested today.

It takes creative skill and thinking to uncover something that is supposed to go undiscovered. Fascinated by the duplicitous possibilities of camouflage, mimicry and deception, scientists have been passionately searching for new forms of natural deception, and expanding the limits of what we think the natural world is capable. In this book I have limited myself to describing cases of deception which are backed by solid evidence or are generally accepted to be plausible cases of deception, even if the data behind them might be a little preliminary. There are countless other supposed cases of natural deception and trickery that haven't found scientific support, are very tenuous hypotheses or are just a bit too out-there to be taken seriously.

Early naturalists looked at the enormous grey backs of hippopotamuses floating just above the water's surface and wondered whether they might be masquerading as submerged boulders. Some have stared at the stripes on the faces of wild cats and perceived the face of an innocent rabbit and pondered whether this is evolution's version of a prowling wolf in sheep's clothing. You will find many other ideas thrown about online that are stated as if they are established facts, rather than tantalising ideas. Again, we are imperfect creatures, and sometimes our ideas – as great as they are – don't quite get it right on the first attempt.

I haven't, for example, discussed in this book the famous mimic octopus *Thaumoctopus mimicus*. This animal has received internet fame for its remarkable colour- and shape-changing abilities. If some blog posts are to be believed, this allows the species to flexibly mimic somewhere around 15 different types of animals, from venomous sea snakes and flounder, to more obscure models such as brittle stars and colonial tunicates. *T. mimicus* was only discovered in 1998 and since then there has been much excited discussion about the implications of its fantastic shape-changing abilities. Research done since then leans towards flounders as the most likely models for the octopus to be mimicking. A few other octopus species have seen been seen doing similar behaviours where they flatten their bodies and swim close to the sea floor. But beyond qualitative descriptions of this behaviour, there haven't been any experiments to test whether this is true mimicry.

As for the other 14 supposed models, it's anybody's guess as to what's going on there. Mimic octopuses are, without a doubt, fascinating animals and these enticing ideas are within the realms of possibility. It's surprising given the morphing abilities of octopuses that their protective strategies would be restricted to camouflage or masquerade, and not mimicry. Currently though, rigorous evidence is lacking. There are numerous videos and photos online of mimic octopuses taken from just the right angle where, with a bit of imagination, you can see the suggested similarities, but this is a case where I think extraordinary claims require extraordinary evidence. It's entirely possible that the mimic octopus is not the masterful mimic we believe it to be, and there could be alternative explanations for its dynamic shape-changing abilities. I could be wrong and, if so, I can't wait to read the extraordinary research that proves it.

Technology has given us tools to move beyond the limitations of our own senses so that we can test how plants and animals communicate using signals we can't perceive. This is where our next great discoveries in understanding deception in nature are likely to occur. The closer we get to understanding the sensory perspectives of other living things, the more likely we are to uncover the limitations of those senses, and where other creatures will be exploiting those limitations for survival. We can acknowledge that great advances have been made and still recognise that we remain heavily biased towards our human perspective on the world. It is when we start to broaden our perspectives beyond the human that we realise how much we still don't know about deception in nature.

Even with technology that allows us to see through other animals' eyes, or listen with their ears, there remains ever greater uncertainty about what is going on in those animal minds. While it's tempting to think that one day, technology will provide us with a tool to read animal thoughts, this is probably a misguided way to think about the problem. If human lie detectors and psychological experiments have shown us anything, it's that we're not certain what happens inside our own minds when we make decisions. Thinking that there is a clear answer to what happens in other animal minds might be wishful thinking. The mindset of other living things may forever be something that we can imagine and infer, but never truly comprehend.

Knowing that there is so much we don't know, can't see, or are fooled into overlooking makes the world a much bigger place. It fills the world with mystery and reminds us that there will always be things we don't understand and more to discover, no matter how hard we look. There are tigers in the jungle and flowers that walk. The next time you wander through the forest or swim in the ocean marvelling at all the wonderful things you see, I hope you are also filled with wonder, intrigue and inspiration by all that you don't see – and maybe never will.

References

For a more scholarly dive into the world of deception in nature, I recommend Martin Stevens' *Cheats and Deceits: How Animals and Plants Exploit and Mislead* (Oxford University Press, 2016). In this book, I avoid exploring whether deception is a fundamental part of human nature. That is a much more complex and heated conversation, and I am not confident that that my musings on natural history would allow me to give much insight. I suspect Lixing Sun might disagree, given that he has written an entire book about it. To hear his perspective, I recommend *The Liars of Nature and the Nature of Liars* (Princeton University Press, 2023).

At a few points in this book, I mentioned how military camouflage has drawn inspiration from animal camouflage. This has a long history, and it continues to this day. To learn more about this, you can read *Dazzled and Deceived: Mimicry and Camouflage* by Peter Forbes (Yale University Press, 2011).

For a joyful exploration of the beauty and bizarreness of orchids, read John Alcock's *An Enthusiasm for Orchids: Sex and Deception in Plant Evolution* (Oxford University Press, 2005).

The dramatic and salacious lives of cuckoos is much more complex than I could cover in a single chapter here. For a more comprehensive tale, I recommend Nick Davies' book *Cuckoo: Cheating by Nature* (Bloomsbury, 2016).

And if you enjoyed stepping out of your comfort zone and into the strange worlds of unfamiliar animal senses, I recommend reading Ed Yong's epic book *An Immense World: How Animal Senses Reveal the Hidden Realms Around Us* (Random House, 2022).

As I mentioned early in the book, the academic literature surrounding sensory ecology is overflowing with opinion pieces, thought pieces, review articles, meta-analyses and many other articles that say the same thing in ever-so-slightly different ways. For academic purposes, you could argue that these are necessary for teasing apart small-scale yet meaningful differences in biological interactions. It could also be argued that it's mostly arm-waving from a bunch of overenthusiastic boffins splitting hairs that may or may not need splitting. Either way, an exhaustive list of references that went into informing this book is not possible here. Instead, I have kept this reference list mercifully short by listing only articles with major impact in the field, review articles that cover broad concepts and examples of research projects specifically mentioned in the book. If you have read this far and are diving into academic references here, reach out and let me know. I'd love to know how often people use this part of the book. Let's use a secret code, hey? Send me the message *the Amazon Mollys are particularly iridescent this Christmas* to let me know you made it this far.

And just in case you were wondering, no part of this book was written using generative AI software. No lousy plagiarism machine could have crafted such gripping prose as this crafty sod has, uh, crafted. Yes, believe it or not, this entire book is entirely the work of a squishy, organic, generally harmless human being.

Chapter 1

Barbosa, A., Allen, J.J., Mäthger, L.M., and Hanlon, R.T. (2012) Cuttlefish use visual cues to determine arm postures for camouflage. *Proceedings of the Royal Society B: Biological Sciences* 279, 84–90.

Barbosa, A., Mäthger, L.M., Buresch, K.C., Kelly, J., Chubb, C., Chiao, C.-C., and Hanlon, R.T. (2008) Cuttlefish camouflage: the effects of substrate contrast and size in evoking uniform, mottle or disruptive body patterns. *Vision Research* 48, 1242–53.

Cocroft, R.B., and Hambler, K. (1989) Observations on a commensal relationship of the microhylid frog *Chiasmocleis ventrimaculata* and the burrowing theraphosid spider *Xenesthis immanis* in Southeastern Peru. *Biotropica* 21, 2–8.

Cronin, T.W. (2016) Camouflage: being invisible in the open ocean. *Current Biology* 26, R1179–81.

Davis, A.L., Thomas, K.N., Goetz, F.E., Robison, B.H., Johnsen, S., and Osborn, K.J. (2020) Ultra-black camouflage in deep-sea fishes. *Current Biology* 30, 3470–76.

Fennell, J.G., Talas, L., Baddeley, R.J., Cuthill, I.C., and Scott-Samuel, N.E. (2019) Optimizing colour for camouflage and visibility using deep learning: the effects of the environment and the observer's visual system. *Journal of the Royal Society Interface* 16, 20190183.

Hanlon, R.T., Naud, M.-J., Forsythe, J.W., Hall, K., Watson, A.C., and McKechnie, J. (2007) Adaptable night camouflage by cuttlefish. *American Naturalist* 169, 543–51.

Johnsen, S. (2014) Hide and seek in the open sea: pelagic camouflage and visual countermeasures. *Annual Review of Marine Science* 6, 369–92.

Lovell, P.G., Ruxton, G.D., Langridge, K.V., and Spencer, K.A. (2013) Egg-laying substrate selection for optimal camouflage by quail. *Current Biology* 23, 260–64.

Mäthger, L.M., Barbosa, A., Miner, S., and Hanlon, R.T. (2006) Color blindness and contrast perception in cuttlefish (*Sepia officinalis*) determined by a visual sensorimotor assay. *Vision Research* 46, 1746–53.

Messenger, J.B. (2001) Cephalopod chromatophores: neurobiology and natural history. *Biological Reviews* 76, 473–528.

Morey, S.R. (1990) Microhabitat selection and predation in the pacific treefrog, *Pseudacris regilla. Journal of Herpetology* 24, 292–96.

Rosa, R., Lopes, V.M., Guerreiro, M., Bolstad, K., and Xavier, J.C. (2017) Biology and ecology of the world's largest invertebrate, the colossal squid (*Mesonychoteuthis hamiltoni*): a short review. *Polar Biology* 40, 1871–83.

Stevens, M., and Merilaita, S. (2009) Animal camouflage: current issues and new perspectives. *Philosophical Transactions of the Royal Society B: Biological Sciences* 364, 423–27.

Stevens, M., Rong, C.P., and Todd, P.A. (2013) Colour change and camouflage in the horned ghost crab *Ocypode ceratophthalmus. Biological Journal of the Linnean Society* 109, 257–70.

Stuart-Fox, D., Whiting, M.J., and Moussalli, A. (2006) Camouflage and colour change: antipredator responses to bird and snake predators across multiple populations in a dwarf chameleon. *Biological Journal of the Linnean Society* 88, 437–46.

Vandenspiegel, D., Ovaere, A., and Massin, C. (1992) On the association between the crab *Hapalonotus reticulatus* (Crustacea, Brachyura, Eumedonidae) and the sea cucumber *Holothuria (Metriatyla) scabra* (Echinodermata, Holothuridae). *Bulletin of the Royal Belgian Institute of Natural Sciences* 62, 167–77.

Chapter 2

Aguilar De Soto, N., Madsen, P.T., Tyack, P., Arranz, P., Marrero, J., Fais, A., Revelli, E., and Johnson, M. (2012) No shallow talk: cryptic strategy in the vocal communication of Blainville's beaked whales. *Marine Mammal Science* 28, E75–92.

Allen, W.L., Cuthill, I.C., Scott-Samuel, N.E., and Baddeley, R. (2011) Why the leopard got its spots: relating pattern development to ecology in felids. *Proceedings of the Royal Society B: Biological Sciences* 278, 1373–80.

Bracken-Grissom, H.D., DeLeo, D.M., Porter, M.L., Iwanicki, T., Sickles, J., and Frank, T.M. (2020) Light organ photosensitivity in deep-sea shrimp may suggest a novel role in counterillumination. *Scientific Reports* 10, 4485.

Brooker, R.M., Munday, P.L., Chivers, D.P., and Jones, G.P. (2015) You are what you eat: diet-induced chemical crypsis in a coral-feeding reef fish. *Proceedings of the Royal Society B: Biological Sciences* 282, 20141887.

Claes, J.M., and Mallefet, J. (2008) Early development of bioluminescence suggests camouflage by counter-illumination in the velvet belly lantern shark *Etmopterus spinax* (Squaloidea: Etmopteridae). *Journal of Fish Biology* 73, 1337–50.

Claes, J.M., and Mallefet, J. (2010) The lantern shark's light switch: turning shallow water crypsis into midwater camouflage. *Biology Letters* 6, 685–87.

Caro, T. (2009) Contrasting coloration in terrestrial mammals. *Philosophical Transactions of the Royal Society B: Biological Sciences* 364, 537–48.

Caro, T., Izzo, A., Reiner, R.C., Walker, H., and Stankowich, T. (2014) The function of zebra stripes. *Nature Communications* 5, 3535.

Cott, H.B. (1940) *Adaptive Coloration in Animals*, Methuen.

Cuthill, I.C., Stevens, M., Sheppard, J., Maddocks, T., Párraga, C.A., and Troscianko, T.S. (2005) Disruptive coloration and background pattern matching. *Nature* 434, 72–74.

Dance, A. (2016) Prehistoric animals, in living color. *Proceedings of the National Academy of Sciences* 113, 8552–56.

Darwin, C. (1859) *On the Origin of Species By Means of Natural Selection*, John Murray.

Davis, A.L., Sutton, T.T., Kier, W.M., and Johnsen, S. (2020) Evidence that eye-facing photophores serve as a reference for counterillumination in an order of deep-sea fishes. *Proceedings of the Royal Society B: Biological Sciences* 287, 20192918.

Ferguson, G.P., Messenger, J.B., and Budelmann, B.U. (1994) Gravity and light influence the countershading reflexes of the cuttlefish *Sepia officinalis*. *Journal of Experimental Biology* 191, 247–56.

Jones, B.W., and Nishiguchi, M.K. (2004) Counterillumination in the Hawaiian bobtail squid, *Euprymna scolopes* Berry (Mollusca: Cephalopoda). *Marine Biology* 144, 1151–55.

Josef, N. (2017) Peri-ocular eye patterning (POEP): more than meets the eye. *Open Journal of Animal Sciences* 7, 356–63.

Kojima, T., Oishi, K., Matsubara, Y., Uchiyama, Y., Fukushima, Y., Aoki, N., Sato, S., Masuda, T., Ueda, J., Hirooka, H., et al. (2019) Cows painted with zebra-like striping can avoid biting fly attack. *PLOS ONE* 14, e0223447.

Mark, C.J., O'Hanlon, J.C., and Holwell, G.I. (2022) Camouflage in lichen moths: field predation experiments and avian vision modelling demonstrate the importance of wing pattern elements and background for survival. *Journal of Animal Ecology* 91, 2358–69.

Massuda, K.F., and Trigo, J.R. (2014) Hiding in plain sight: cuticular compound profile matching conceals a larval tortoise beetle in its host chemical cloud. *Journal of Chemical Ecology* 40, 341–54.

Melin, A.D., Kline, D.W., Hiramatsu, C., and Caro, T. (2016) Zebra stripes through the eyes of their predators, zebras, and humans. *PLOS ONE* 11, e0145679.

Miller, A.K., Maritz, B., McKay, S., Glaudas, X., and Alexander, G.J. (2015) An ambusher's arsenal: chemical crypsis in the puff adder (*Bitis arietans*). *Proceedings of the Royal Society B: Biological Sciences* 282, 20152182.

Morisaka, T., and Connor, R.C. (2007) Predation by killer whales (*Orcinus orca*) and the evolution of whistle loss and narrow-band high frequency clicks in odontocetes. *Journal of Evolutionary Biology* 20, 1439–58.

Rowland, H.M. (2009) From Abbott Thayer to the present day: what have we learned about the function of countershading? *Philosophical Transactions of the Royal Society B: Biological Sciences* 364, 519–27.

Ruxton, G.D. (2009) Non-visual crypsis: a review of the empirical evidence for camouflage to senses other than vision. *Philosophical Transactions of the Royal Society B: Biological Sciences* 364, 549–57.

Schaefer, H.M., and Stobbe, N. (2006) Disruptive coloration provides camouflage independent of background matching. *Proceedings of the Royal Society B: Biological Sciences* 273, 2427–32.

Smithwick, F.M., Nicholls, R., Cuthill, I.C., and Vinther, J. (2017) Countershading and stripes in the theropod dinosaur *Sinosauropteryx* reveal heterogeneous habitats in the Early Cretaceous Jehol Biota. *Current Biology* 27, 3337–43.

Takács, P., Száz, D., Vincze, M., Slíz-Balogh, J., and Horváth, G. (2022) Sunlit zebra stripes may confuse the thermal perception of blood vessels causing the visual unattractiveness of zebras to horseflies. *Scientific Reports* 12, 10871.

Thayer, A. (1909) *Concealing-Coloration in the Animal Kingdom*, Macmillan.

Turunen, S., Paavilainen, S., Vepsäläinen, J., and Hielm-Björkman, A. (2024) Scent detection threshold of trained dogs to *Eucalyptus* hydrolat. *Animals* 14, 1083.

Vinther, J. (2015) A guide to the field of palaeo colour: melanin and other pigments can fossilise: Reconstructing colour patterns from ancient organisms can give new insights to ecology and behaviour. *BioEssays* 37, 643–56.

Vinther, J., Nicholls, R., Lautenschlager, S., Pittman, M., Kaye, T.G., Rayfield, E., Mayr, G., and Cuthill, I.C. (2016) 3D camouflage in an ornithischian dinosaur. *Current Biology* 26, 2456–62.

Chapter 3

Akino, T., Nakamura, K.-I., and Wakamura, S. (2004) Diet-induced chemical phytomimesis by twig-like caterpillars of *Biston robustum* Butler (Lepidoptera: Geometridae). *Chemoecology* 14, 165–74.

Barlow, B.A., and Wiens, D. (1977) Host-parasite resemblance in Australian mistletoes: the case for cryptic mimicry. *Evolution* 31, 69–84.

Bates, H.W. (1862) XXXII. Contributions to an insect fauna of the Amazon valley. Lepidoptera: Heliconidæ. *Transactions of the Linnean Society of London* 3, 495–566.

Bian, X., Elgar, M.A., and Peters, R.A. (2016) The swaying behavior of *Extatosoma tiaratum*: motion camouflage in a stick insect? *Behavioral Ecology* 27, 83–92.

Blick, R.A.J., Burns, K.C., and Moles, A.T. (2012) Predicting network topology of mistletoe–host interactions: do mistletoes really mimic their hosts? *Oikos* 121, 761–71.

Darwin, F. (ed.) (1887) *The Life and Letters of Charles Darwin*, vol. 2, John Murray.

Dowdy, N.J., and Conner, W.E. (2016) Acoustic aposematism and evasive action in select chemically defended Arctiine (Lepidoptera: Erebidae) species: nonchalant or not? *PLOS ONE* 11, e0152981.

Fritz, M. (1879) Ituna and Thyridia: a remarkable case of mimicry in butterflies. *Transactions of the Entomological Society of London* 1879, 20–29.

Gianoli, E. (2017) Eyes in the chameleon vine? *Trends in Plant Science* 22, 4–5.

Gianoli, E., and Carrasco-Urra, F. (2014) Leaf mimicry in a climbing plant protects against herbivory. *Current Biology* 24, 984–87.

Gianoli, E., González-Teuber, M., Vilo, C., Guevara-Araya, M.J., and Escobedo, V.M. (2021) Endophytic bacterial communities are associated with leaf mimicry in the vine *Boquila trifoliolata*. *Scientific Reports* 11, 22673.

Ito, F., Hashim, R., Huei, Y.S., Kaufmann, E., Akino, T., and Billen, J. (2004). Spectacular Batesian mimicry in ants. *Naturwissenschaften* 91, 481–84.

Klomp, D.A., Stuart-Fox, D., Das, I., and Ord, T.J. (2014) Marked colour divergence in the gliding membranes of a tropical lizard mirrors population differences in the colour of falling leaves. *Biology Letters* 10, 20140776.

Kraemer, A.C., and Adams, D.C. (2014) Predator perception of Batesian mimicry and conspicuousness in a salamander: brief communication. *Evolution* 68, 1197–1206.

Lev-Yadun, S. (2002) Defensive ant, aphid and caterpillar mimicry in plants? *Biological Journal of the Linnean Society* 77, 393–98.

—— (2009) Ant mimicry by *Passiflora* flowers? *Israel Journal of Entomology* 39, 159–63.

—— (2009) Müllerian and Batesian mimicry rings of white-variegated aposematic spiny and thorny plants: A hypothesis. *Israel Journal of Plant Sciences* 57, 107–16.

—— (2013) The enigmatic fast leaflet rotation in *Desmodium motorium*: Butterfly mimicry for defense? *Plant Signaling and Behavior* 8, e24473.

—— (2014) Defensive masquerade by plants. *Biological journal of the Linnean Society* 113, 1162–66.

McLean, D.J., Cassis, G., and Herberstein, M.E. (2024) Morphological ant mimics: constrained to imperfection? *Biology Letters* 20, 20230330.

Merrill, R.M., Dasmahapatra, K.K., Davey, J.W., Dell'Aglio, D.D., Hanly, J.J., Huber, B., Jiggins, C.D., Joron, M., Kozak, K.M., Llaurens, V., et al. (2015) The diversification of *Heliconius* butterflies: what have we learned in 150 years? *Journal of Evolutionary Biology* 28, 1417–38.

Pekár, S., Petráková, L., Bulbert, M.W., Whiting, M.J., and Herberstein, M.E. (2017) The golden mimicry complex uses a wide spectrum of defence to deter a community of predators. *eLife* 6, e22089.

Rubino, D.L., and McCarthy, B.C. (2004) Presence of aposematic (warning) coloration in vascular plants of southeastern Ohio. *Journal of the Torrey Botanical Society* 131, 252–56.

Sena, A.T., and Ruane, S. (2022) Concepts and contentions of coral snake resemblance: Batesian mimicry and its alternatives. *Biological Journal of the Linnean Society* 135, 631–44.

Skelhorn, J., Rowland, H.M., Speed, M.P., and Ruxton, G.D. (2010) Masquerade: camouflage without crypsis. *Science* 327, 51.

Soltau, U., Dötterl, S., and Liede-Schumann, S. (2009) Leaf variegation in *Caladium steudneriifolium* (Araceae): a case of mimicry? *Evolutionary Ecology* 23, 503–12.

Stuckert, A.M.M., and Summers, K. (2023) Investigating signal modalities of aposematism in a poison frog. *Journal of Evolutionary Biology* 36, 1003–1009.

Wallace, A.R. (1877) The colours of animals and plants. *Macmillan's Magazine* 36, 384–408.

White, J., and Yamashita, F. (2022) *Boquila trifoliolata* mimics leaves of an artificial plastic host plant. *Plant Signaling and Behavior* 17, 1977530.

Zhang, Z.-T., Yu, L., Chang, H.-Z., Zhang, S.-C., and Li, D.-Q (2024) Nature's disguise: empirical demonstration of dead-leaf masquerade in *Kallima* butterflies. *Zoological Research* 45, 1201–1208.

Chapter 4

Averill-Murray, R.C. (2006) Natural history of the western hog-nosed snake (*Heterodon nasicus*) with notes on envenomation. *Sonoran Herpetologist* 19, 98–101.

Badiane, A., Carazo, P., Price-Rees, S.J., Ferrando-Bernal, M., and Whiting, M.J. (2018) Why blue tongue? A potential UV-based deimatic display in a lizard. *Behavioral Ecology and Sociobiology* 72, 104.

Barber, J.R., Leavell, B.C., Keener, A.L., Breinholt, J.W., Chadwell, B.A., McClure, C.J.W., Hill, G.M., and Kawahara, A.Y. (2015) Moth tails divert bat attack: evolution of acoustic deflection. *Proceedings of the National Academy of Sciences* 112, 2812–16.

Behrens, R.R. (1999) The role of artists in ship camouflage during World War I. *Leonardo* 32, 53–59.

Chotard, A., Ledamoisel, J., Decamps, T., Herrel, A., Chaine, A.S., Llaurens, V., and Debat, V. (2002) Evidence of attack deflection suggests adaptive evolution of wing tails in butterflies. *Proceedings of the Royal Society B: Biological Sciences* 289, 20220562.

Conway, B.R. (2005) Neural basis for a powerful static motion illusion. *Journal of Neuroscience* 25, 5651–56.

Hart, N.S., and Collin, S.P. (2015) Sharks senses and shark repellents. *Integrative Zoology* 10, 38–64.

Hendrick, L.K., Somjee, U., Rubin, J.J., and Kawahara, A.Y. (2022) A review of false heads in lycaenid butterflies. *Journal of the Lepidopterists' Society* 76, 140–48.

Humphreys, R.K., and Ruxton, G.D. (2018) A review of thanatosis (death feigning) as an anti-predator behaviour. *Behavioral Ecology and Sociobiology* 72, 1–16.

Kitaoka, A., and Ashida, H. (2003) Phenomenal characteristics of the peripheral drift illusion. *Vision* 15, 261–62.

Kjernsmo, K., Grönholm, M., and Merilaita, S. (2016) Adaptive constellations of protective marks: eyespots, eye stripes and diversion of attacks by fish. *Animal Behaviour* 111, 189–95.

Lafitte, A., Sordello, R., Legrand, M., Nicolas, V., Obein, G., and Reyjol, Y. (2022) A flashing light may not be that flashy: a systematic review on critical fusion frequencies. *PLOS ONE* 17, e0279718.

Langridge, K.V., Broom, M., and Osorio, D. (2007) Selective signalling by cuttlefish to predators. *Current Biology* 17, R1044–45.

Lee, W.-J., and Moss, C.F. (2016) Can the elongated hindwing tails of fluttering moths serve as false sonar targets to divert bat attacks? *Journal of the Acoustical Society of America* 139, 2579–88.

Lindell, L.E., and Forsman, A. (1996) Sexual dichromatism in snakes: support for the flicker-fusion hypothesis. *Canadian Journal of Zoology* 74, 2254–56.

O'Hanlon, J.C., Rathnayake, D.N., Barry, K.L., and Umbers, K.D.L. (2018) Post-attack defensive displays in three praying mantis species. *Behavioral Ecology and Sociobiology* 72, 176.

Olofsson, M., Løvlie, H., Tibblin, J., Jakobsson, S., and Wiklund, C. (2013) Eyespot display in the peacock butterfly triggers antipredator behaviors in naïve adult fowl. *Behavioral Ecology* 24, 305–10.

Perez-Martinez, C.A., Riley, J.L., and Whiting, M.J. (2020) Uncovering the function of an enigmatic display: antipredator behaviour in the iconic Australian frillneck lizard. *Biological Journal of the Linnean Society* 129, 425–38.

Rasmussen, A.R., and Elmberg, J. (2009) 'Head for my tail': a new hypothesis to explain how venomous sea snakes avoid becoming prey. *Marine Ecology* 30, 385–90.

Ryan, L.A., Gennari, E., Slip, D.J., Collin, S.P., Peddemors, V.M., Huveneers, C., Chapuis, L., Hemmi, J., and Hart, N.S. (2024) Counterillumination reduces bites by Great White sharks. *Current Biology* 34, 5789–95.

Stevens, M. (2005) The role of eyespots as anti-predator mechanisms, principally demonstrated in the Lepidoptera. *Biological Reviews* 80, 573–88.

Titcomb, G.C., Kikuchi, D.W., and Pfennig, D.W. (2014) More than mimicry? Evaluating scope for flicker-fusion as a defensive strategy in coral snake mimics. *Current Zoology* 60, 123–30.

Umbers, K.D.L., and Mappes, J. (2015) Postattack deimatic display in the mountain katydid, *Acripeza reticulata. Animal Behaviour* 100, 68–73.

Umeton, D., Tarawneh, G., Fezza, E., Read, J.C.A., and Rowe, C. (2019) Pattern and speed interact to hide moving prey. *Current Biology* 29, 3109–13.

Vallin, A., Jakobsson, S., Lind, J., and Wiklund, C. (2005) Prey survival by predator intimidation: an experimental study of peacock butterfly defence against blue tits. *Proceedings of the Royal Society B: Biological Sciences* 272, 1203–07.

Vidal-García, M., O'Hanlon, J.C., Svenson, G.J., and Umbers, K.D.L. (2020) The evolution of startle displays: a case study in praying mantises. *Proceedings of the Royal Society B: Biological Sciences* 287, 20201016.

Watson, C.M., Roelke, C.E., Pasichnyk, P.N., and Cox, C.L. (2012) The fitness consequences of the autotomous blue tail in lizards: an empirical test of predator response using clay models. *Zoology* 115, 339–44.

Zanker, J.M., and Walker, R. (2004) A new look at Op art: towards a simple explanation of illusory motion. *Naturwissenschaften* 91, 149–56.

Chapter 5

Bulbert, M.W., Herberstein, M.E., and Cassis, G. (2014) Assassin bug requires dangerous ant prey to bite first. *Current Biology* 24, R220–21.

Cheney, K.L., and Côté, I.M. (2005) Frequency-dependent success of aggressive mimics in a cleaning symbiosis. *Proceedings of the Royal Society B: Biological Sciences* 272, 2635–39.

da Fonseca, W., Correa, R.R., de Souza Oliveira, A., and Bernarde, P.S. (2019) Caudal luring in the Neotropical two-striped forest pitviper *Bothrops bilineatus smaragdinus* Hoge, 1966 in the western Amazon. *Herpetology Notes* 12, 365–74.

Drummond, H., and Gordon, E.R. (2010) Luring in the neonate alligator snapping turtle (*Macroclemys temminckii*): description and experimental analysis. *Zeitschrift für Tierpsychologie* 50, 136–52.

Eberhard, W.G. (1980) The natural history and behavior of the bolas spider *Mastophora dizzydeani* sp. n. (Araneidae). *Psyche* 87, 143–69.

Fathinia, B., Rastegar-Pouyani, N., Rastegar-Pouyani, E., Todehdehghan, F., and Amiri, F. (2015) Avian deception using an elaborate caudal lure in *Pseudocerastes urarachnoides* (Serpentes: Viperidae). *Amphibia-Reptilia* 36, 223–31.

Haddock, S.H.D., Dunn, C.W., Pugh, P.R., and Schnitzler, C.E. (2005) Bioluminescent and red-fluorescent lures in a deep-sea siphonophore. *Science* 309, 263.

Hansknecht, K.A. (2008) Lingual luring by mangrove saltmarsh snakes (*Nerodia clarkii compressicauda*). *Journal of Herpetology* 42, 9–15.

Heiling, A.M., Herberstein, M.E., and Chittka, L. (2003) Crab-spiders manipulate flower signals. *Nature* 421, 3345.

Hingston, J. (1879) *The Australian Abroad: Branches from the main route round the world*, Sampson Low, Marston, Searle and Rivington.

Jacobson, E. (1911) Biological notes on the Hemipteron *Ptilocerus ochraceus*. *Tijdschrift Voor Entomologie* 54, 175–79.

Lloyd, J.E. (1975) Aggressive mimicry in *Photuris* fireflies: signal repertoires by femmes fatales. *Science* 187, 452–53.

Marshall, D.C., and Hill, K.B.R. (2009) Versatile aggressive mimicry of cicadas by an Australian predatory katydid. *PLOS ONE* 4, e4185.

Master, T.L. (1991) Use of tongue-flicking behaviour by the snowy egret. *Journal of Field Ornithology* 62, 399–402.

O'Hanlon, J.C. (2014) The roles of colour and shape in pollinator deception in the orchid mantis *Hymenopus coronatus*. *Ethology* 120, 652–61.

O'Hanlon, J.C., Holwell, G.I., and Herberstein, M.E. (2014) Pollinator deception in the orchid mantis. *American Naturalist* 183, 126–32.

Schaefer, H.M., and Ruxton, G.D. (2008) Fatal attraction: carnivorous plants roll out the red carpet to lure insects. *Biology Letters* 4, 153–55.

Weirauch, C., Bulbert, M., and Cassis, G. (2010) Comparative trichome morphology in feather-legged assassin bugs (Insecta: Heteroptera: Reduviidae: Holoptilinae). *Zoologischer Anzeiger-A Journal of Comparative Zoology* 248, 237–53.

Welsh, H.H., and Lind, A.J. (2000) Evidence of lingual-luring by an aquatic snake. *Journal of Herpetology* 34, 67–74.

White, T.E. (2017) Jewelled spiders manipulate colour-lure geometry to deceive prey. *Biology Letters* 13, 20170027.

Wignall, A.E., and Taylor, P.W. (2011) Assassin bug uses aggressive mimicry to lure spider prey. *Proceedings of the Royal Society B: Biological Sciences* 278, 1427–33.

Willis, R.E., White, C.R., and Merritt, D.J. (2010) Using light as a lure is an efficient predatory strategy in *Arachnocampa flava*, an Australian glowworm. *Journal of Comparative Physiology B* 181, 477–86.

Yu, L., Xu, X., Zhang, Z., Painting, C.J., Yang, X., and Li, D. (2021) Masquerading predators deceive prey by aggressively mimicking bird droppings in a crab spider. *Current Zoology* 68, 325–34.

Chapter 6

Angioy, A.-M., Stensmyr, M.C., Urru, I., Puliafito, M., Collu, I., and Hansson, B.S. (2004) Function of the heater: the dead horse arum revisited. *Proceedings of the Royal Society B: Biological Sciences* 271. S13–15.

Bilde, T., Tuni, C., Elsayed, R., Pekár, S., and Toft, S. (2006) Death feigning in the face of sexual cannibalism. *Biology Letters* 2, 23–25.

Brodmann, J., Twele, R., Francke, W., Yi-bo, L., Xi-qiang, S., and Ayasse, M. (2009) Orchid mimics honey bee alarm pheromone in order to attract hornets for pollination. *Current Biology* 19, 1368–72.

Brown, C., Garwood, M.P., and Williamson, J.E. (2012) It pays to cheat: tactical deception in a cephalopod social signalling system. *Biology Letters* 8, 729–32.

Brunton Martin, A.L., Gaskett, A.C., and O'Hanlon, J.C. (2021) Museum records indicate male bias in pollinators of sexually deceptive orchids. *The Science of Nature* 108, 25.

Brunton Martin, A.L., O'Hanlon, J.C., and Gaskett, A.C. (2020) Orchid sexual deceit affects pollinator sperm transfer. *Functional Ecology* 34, 1336–44.

Dries, L.A. (2003) Peering through the looking glass at a sexual parasite: are amazon mollies red queens? *Evolution* 57, 1387–96.

Gaskett, A.C., Winnick, C.G., and Herberstein, M.E. (2008) Orchid sexual deceit provokes ejaculation. *American Naturalist* 171, E206–12.

Ghislandi, P.G., Albo, M.J., Tuni, C., and Bilde, T. (2014) Evolution of deceit by worthless donations in a nuptial gift-giving spider. *Current Zoology* 60, 43–51.

Jersáková, J., Spaethe, J., Streinzer, M., Neumayer, J., Paulus, H., Dötterl, S., and Johnson, S.D. (2016) Does *Traunsteinera globosa* (the globe orchid) dupe its pollinators through generalized food deception or mimicry? *Botanical Journal of the Linnean Society* 180, 269–94.

Johnson, S.D., and Morita, S. (2006) Lying to Pinocchio: floral deception in an orchid pollinated by long-proboscid flies. *Botanical Journal of the Linnean Society* 152, 271–78.

Jukema, J., and Piersma, T. (2006) Permanent female mimics in a lekking shorebird. *Biology Letters* 2, 161–64.

Khelifa, R. (2017) Faking death to avoid male coercion: extreme sexual conflict resolution in a dragonfly. *Ecology* 98, 1724–26.

Lobin, W., Neumann, M., Radscheit, M., and Barthlott, W. (2007) The cultivation of titan arum *Amorphophallus titanum* – a flagship species for botanic gardens. *Sibbaldia: the International Journal of Botanic Garden Horticulture* 5, 69–86.

Nikolov, L.A., and Davis, C.C. (2017) The big, the bad, and the beautiful: biology of the world's largest flowers. *Journal of Systematics and Evolution* 55, 516–24.

Norman, M.D., Finn, J., and Tregenza, T. (1999) Female impersonation as an alternative reproductive strategy in giant cuttlefish. *Proceedings of the Royal Society B: Biological Sciences* 266, 1347–49.

Porter, B., Fiumera, A., and Avise, J. (2002) Egg mimicry and allopaternal care: two mate-attracting tactics by which nesting striped darter (*Etheostoma virgatum*) males enhance reproductive success. *Behavioral Ecology and Sociobiology* 51, 350–59.

Roy, B. (1999) Floral mimicry: A fascinating yet poorly understood phenomenon. *Trends in Plant Science* 4, 325–30.

Schlupp, I., Riesch, R., and Tobler, M. (2007) Amazon mollies. *Current Biology* 17, R536–37.

Spaethe, J., Moser, W.H., and Paulus, H.F. (2007) Increase of pollinator attraction by means of a visual signal in the sexually deceptive orchid, *Ophrys heldreichii* (Orchidaceae). *Plant Systematics and Evolution* 264, 31–40.

Stökl, J., Brodmann, J., Dafni, A., Ayasse, M., and Hansson, B.S. (2011) Smells like aphids: orchid flowers mimic aphid alarm pheromones to attract hoverflies for pollination. *Proceedings of the Royal Society B: Biological Sciences* 278, 1216–22.

Suetsugu, K. (2022) *Arisaema:* pollination by lethal attraction. *Plants People Planet* 4, 196–200.

Suetsugu, K., Nishigaki, H., Kakishima, S., Sueyoshi, M., and Sugiura, S. (2024) Back from the dead: a fungus gnat pollinator turns *Arisaema* lethal trap into nursery. *Plants People Planet* 6, 536–43.

Urru, I., Stensmyr, M.C., and Hansson, B.S. (2011) Pollination by brood-site deception. *Phytochemistry* 72, 1655–66.

Whiting, M.J., Webb, J.K., and Keogh, J.S. (2009) Flat lizard female mimics use sexual deception in visual but not chemical signals. *Proceedings of the Royal Society B: Biological Sciences* 276, 1585–91.

Chapter 7
Brown, C.R., and Bomberger Brown, M. (1989) Behavioural dynamics of intraspecific brood parasitism in colonial cliff swallows. *Animal Behaviour* 37, 777–96.

De Mársico, M.C., Gantchoff, M.G., and Reboreda, J.C. (2012) Host–parasite coevolution beyond the nestling stage? Mimicry of host fledglings by the specialist screaming cowbird. *Proceedings of the Royal Society B: Biological Sciences* 279, 3401–08.

Feeney, W.E., Troscianko, J., Langmore, N.E., and Spottiswoode, C.N. (2015) Evidence for aggressive mimicry in an adult brood parasitic bird, and generalized defences in its host. *Proceedings of the Royal Society B: Biological Sciences* 282, 20150795.

Hafernik, J., and Saul-Gershenz, L. (2000) Beetle larvae cooperate to mimic bees. *Nature* 405, 35–36.

Hayes, M.P. (2015) The biology and ecology of the large blue butterfly *Phengaris* (Maculinea) *arion*: a review. *Journal of Insect Conservation* 19, 1037–51.

Langmore, N.E., Maurer, G., Adcock, G.J., and Kilner, R.M. (2008) Socially acquired host-specific mimicry and the evolution of host races in Horsfield's bronze-cuckoo *Chalcites basalis*. *Evolution* 62, 1689–99.

Matsuura, K. (2006) Termite-egg mimicry by a sclerotium-forming fungus. *Proceedings of the Royal Society B: Biological Sciences* 273, 1203–09.

Rojas Ripari, J.M., Ursino, C.A., Reboreda, J.C., and De Mársico, M.C. (2021) Tricking parents: a review of mechanisms and signals of host manipulation by brood-parasitic young. *Frontiers of Ecology and Evolution* 9, 725792.

Stevens, M. (2013) Bird brood parasitism. *Current Biology* 23, R909–13.

Strohm, E., Kroiss, J., Herzner, G., Laurien-Kehnen, C., Boland, W., Schreier, P., and Schmitt, T. (2008) A cuckoo in wolves' clothing? Chemical mimicry in a specialized cuckoo wasp of the European beewolf (Hymenoptera, Chrysididae and Crabronidae). *Frontiers in Zoology* 5, 2.

Tanaka, K.D., and Ueda, K. (2005) Horsfield's hawk-cuckoo nestlings simulate multiple gapes for begging. *Science* 308, 653–53.

Wang, N., and Kimball, R.T. (2012) Nestmate killing by obligate brood parasitic chicks: is this linked to obligate siblicidal behavior? *Journal of Ornithology* 153, 825–31.

York, J.E. (2021) The evolution of predator resemblance in avian brood parasites. *Frontiers in Ecology and Evolution* 9, 725842.

Chapters 8 and 9

Bond, C.F., and DePaulo, B.M. (2006) Accuracy of deception judgments. *Personality and Social Psychology Review* 10, 214–34.

Byrne, R.W., and Corp, N. (2004) Neocortex size predicts deception rate in primates. *Proceedings of the Royal Society B: Biological Sciences* 271, 1693–99.

Cook, L., and Mitschow, L. (2019) Beyond the polygraph: deception detection and the autonomic nervous system. *Federal Practitioner* 36, 316.

Flower, T.P., Gribble, M., and Ridley, A.R. (2014) Deception by flexible alarm mimicry in an African bird. *Science* 344, 513–16.

Hall, K., and Brosnan, S.F. (2017) Cooperation and deception in primates. *Infant Behavior and Development* 48, 38–44.

Heberlein, M.T.E., Manser, M.B., and Turner, D.C. (2017) Deceptive-like behaviour in dogs (*Canis familiaris*). *Animal Cognition* 20, 511–20.

Møller, A.P. (1988) False alarm calls as a means of resource usurpation in the great tit *Parus major*. *Ethology* 79, 25–30.

Norman, M.D., Finn, J., and Tregenza, T. (2001) Dynamic mimicry in an Indo–Malayan octopus. *Proceedings of the Royal Society B: Biological Sciences* 268, 1755–58.

Owen, R. (1866) *On the Anatomy of Vertebrates*. Longmans, Green and Co.

Spence, S.A., Farrow, T.F.D., Herford, A.E., Wilkinson, I.D., Zheng, Y., and Woodruff, P.W.R. (2001) Behavioural and functional anatomical correlates of deception in humans. *Neuroreport* 12, 2849–53.

Steele, M.A., Halkin, S.L., Smallwood, P.D., McKenna, T.J., Mitsopoulos, K., and Beam, M. (2008) Cache protection strategies of a scatter-hoarding rodent: do tree squirrels engage in behavioural deception? *Animal Behaviour* 75, 705–14.

Vicianova, M. (2015) Historical techniques of lie detection. *Europe's Journal of Psychology* 11, 522–34.

Wang, T., Kong, Y., Zhang, H., Li, Y., Hou, R., Dunn, D.W., Hou, X., Huang, K., and Li, B. (2023) Do golden snub-nosed monkeys use deceptive alarm calls during competition for food? *iScience* 26, 106098.

Woodruff, G., and Premack, D. (1979) Intentional communication in the chimpanzee: The development of deception. *Cognition* 7, 333–62.

Acknowledgements

There are many biology rockstars that have been name-dropped in this book, and there are many more that have made the groundbreaking discoveries that informed the stories told here. To name a few: Innes Cuthill, Martin Stevens, Hannah Rowland, John Endler and Malcolm Edmunds are (in no particular order) the David Bowies, Taylor Swifts, Simon & Garfunkels and Milli Vanillis of sensory ecology research. I have had the pleasure of collaborating with many others where we uncovered more weird and wonderful ways that animals and plants deceive. A big shout-out to Thomas White, Anne Gaskett, Kate Umbers, Gregory Holwell, Marie Herberstein, Matthew Bulbert and Daiqin Li. I will let each of you decide which musical counterpart you wish to adopt in celebration of your contributions to the field. Thanks also to Cassandra Mark-Chan, Amy Brunton Martin, Michael Whitehead, Martin Garwood and Jim McClean for being both amazing scientists and brilliant photographers whose work you can find on display in this book.

Once again thanks to Harriet McInerney and the team at NewSouth Publishing for putting their faith in my abilities and bringing this book into existence. An enormous thanks goes to Siobhan, Mara and Edie for both distracting me from this book and pushing me onwards to get it done. Finally, to Marian O'Hanlon, thank you for your years of trust and patience when I said I was jumping on planes to meet strangers that I met on the internet who promised to show me insects. Mum, this book is for you.